RAND NATIONAL DEFENSE RESEARCH INSTITUTE

T0146370

SEXUAL ASSAULT AND SEXUAL HARASSMENT IN THE U.S. MILITARY

Volume 3. Estimates for Coast Guard Service Members from the 2014 RAND Military Workplace Study

Andrew R. Morral, Kristie L. Gore, Terry L. Schell, editors

Prepared for the DoD Sexual Assault Prevention and Response Office

For more information on this publication, visit www.rand.org/t/RR870z4

Library of Congress Cataloging-in-Publication Data is available for this publication.
ISBN: 978-0-8330-9054-6

Support RAND
Make a tax-deductible charitable contribution at
www.rand.org/giving/contribute

www.rand.org

The 2014 RAND Military Workplace Study Team

Principal Investigators
Andrew R. Morral, Ph.D.
Kristie L. Gore, Ph.D.

Instrument Design
Lisa Jaycox, Ph.D., team lead
Terry Schell, Ph.D.
Coreen Farris, Ph.D.
Dean Kilpatrick, Ph.D.*
Amy Street, Ph.D.*
Terri Tanielian, M.A.*

Study Design and Analysis
Terry Schell, Ph.D., team lead
Bonnie Ghosh-Dastidar, Ph.D.
Craig Martin, M.A.
Q Burkhart, M.S.
Robin Beckman, M.P.H.
Megan Mathews, M.A.
Marc Elliott, Ph.D.

Project Management
Kayla M. Williams, M.A.
Caroline Epley, M.P.A.
Amy Grace Donohue, M.P.P.

Survey Coordination
Jennifer Hawes-Dawson

Westat Survey Group
Shelley Perry, Ph.D., team lead
Wayne Hintze, M.S.
John Rauch
Bryan Davis
Lena Watkins
Richard Sigman, M.S.
Michael Hornbostel, M.S.

Project Communications
Steve Kistler
Jeffrey Hiday
Barbara Bicksler, M.P.P.

Scientific Advisory Board

Major General John Altenburg, Esq. (USA, ret.)
Captain Thomas A. Grieger, M.D. (USN, ret.)
Dean Kilpatrick, Ph.D.
Laura Miller, Ph.D.
Amy Street, Ph.D.
Roger Tourangeau, Ph.D.

David Cantor, Ph.D.
Colonel Dawn Hankins, USAF
Roderick Little, Ph.D.
Sharon Smith, Ph.D.
Terri Tanielian, M.A.
Veronica Venture, J.D.

* Three members of the Scientific Advisory Board were so extensively involved in the development of the survey instrument that we list them here as full Instrument Design team members.

Preface

The Department of Defense (DoD) has assessed service member experiences with sexual assault and harassment since at least 1996, when Public Law 104-201 first required a survey of the "gender relations climate" experienced by active-component forces. Since 2002, four "Workplace and Gender Relations Surveys," as they are known in 10 U.S.C. §481, have been conducted with active-component forces (in 2002, 2006, 2010, and 2012). DoD conducted reserve-component versions of this survey in 2004, 2008, and 2012.

The results of the 2012 survey suggested that more than 26,000 service members in the active component had experienced *unwanted sexual contacts* in the prior year, an estimate that received widespread public attention and concern. In press reports and congressional inquiries, questions were raised about the validity of the estimate, about what "unwanted sexual contact" included, and about whether the survey had been conducted properly. Because of these questions, some members of Congress urged DoD to seek an independent assessment of the number of service members who experienced sexual assault or sexual harassment.

The Sexual Assault Prevention and Response Office within the Office of the Secretary of Defense selected the RAND Corporation to provide a new and independent evaluation of sexual assault, sexual harassment, and gender discrimination across the military. As such, DoD asked the RAND research team to redesign the approach used in previous DoD surveys, if changes would improve the accuracy and validity of the survey results for estimating the prevalence of sexual crimes and violations. In the summer of 2014, RAND fielded a new survey as part of the RAND Military Workplace Study.

This report, Volume 3 in our series, presents survey results for the U.S. Coast Guard and the Coast Guard Reserve. The complete series that collectively describes the study methodology and its findings includes the following reports:

- *Sexual Assault and Sexual Harassment in the U.S. Military: Top-Line Estimates for Active-Duty Service Members from the 2014 RAND Military Workplace Study*
- *Sexual Assault and Sexual Harassment in the U.S. Military: Top-Line Estimates for Active-Duty Coast Guard Members from the 2014 RAND Military Workplace Study*

- *Sexual Assault and Sexual Harassment in the U.S. Military: Volume 1. Design of the 2014 RAND Military Workplace Study*
- *Sexual Assault and Sexual Harassment in the U.S. Military: Volume 2. Estimates for Department of Defense Service Members from the 2014 RAND Military Workplace Study*
- *Sexual Assault and Sexual Harassment in the U.S. Military: Annex to Volume 2. Tabular Results from the 2014 RAND Military Workplace Study for Department of Defense Service Members*
- *Sexual Assault and Sexual Harassment in the U.S. Military: Volume 3. Estimates for Coast Guard Service Members from the 2014 RAND Military Workplace Study*
- *Sexual Assault and Sexual Harassment in the U.S. Military: Annex to Volume 3. Tabular Results from the 2014 RAND Military Workplace Study for Coast Guard Service Members*
- *Sexual Assault and Sexual Harassment in the U.S. Military: Volume 4. Investigations of Potential Bias in Estimates from the 2014 RAND Military Workplace Study.*

These reports are available online at www.rand.org/surveys/rmws.html.

This research was conducted within the Forces and Resources Policy Center of the RAND National Defense Research Institute, a federally funded research and development center sponsored by the Office of the Secretary of Defense, the Joint Staff, the Unified Combatant Commands, the Navy, the Marine Corps, the defense agencies, and the defense Intelligence Community.

For more information on the Forces and Resources Policy Center, see http://www.rand.org/nsrd/ndri/centers/frp.html or contact the director (contact information is provided on the web page).

Contents

Preface v
Figures and Tables ix
Summary xiii
Acknowledgments xix
Abbreviations xxi

CHAPTER ONE

Introduction 1

CHAPTER TWO

Study Design and Analysis Approach 3
Study Design and Sample 3
Statistical Analysis and Reporting Conventions Used in This Report 6

CHAPTER THREE

Sexual Assault Findings: Coast Guard Active Component 9
Sexual Assault Prevalence 9
Unwanted Events and Types of Events Categorized as Past-Year Sexual Assault 13
Reports of Sexual Assaults Prior to the Past Year 15
Characteristics of the Sexual Assault or the "Most Serious" of Multiple Assaults in the
    Past Year 18
Summary 22

CHAPTER FOUR

**Sexual Harassment and Gender Discrimination Findings: Coast Guard Active
    Component** 23
Prevalence of Sexual Harassment and Gender Discrimination 23
Relationship Between Pay Grade and Sexual Harassment 29
Relationship Between Pay Grade and Gender Discrimination 29
Co-Occurrence of Sexual Harassment and Gender Discrimination 30
Inappropriate Workplace Behaviors 30

Types of Sexual Harassment and Gender Discrimination Violations...........................33
Self-Identification of Events as Sexual Harassment 35
Description of Past-Year Sexual Harassment or Gender Discrimination......................37
Summary..44

CHAPTER FIVE
Beliefs About Sexual Assault and Sexual Harassment Prevalence, Prevention, and
   Progress...45
Perceptions of Safety..45
Perceptions of Frequency of Sexual Harassment and Discrimination Against Women 46
Attitudes and Expectations for Justice..47
Likelihood of Reporting Behaviors and Taking Action 47
Perceptions of Unit Leadership...47
Beliefs About Personal Responsibility for Others and Trust in the Military System 48
Perceptions of Progress 48
Perceptions of and Satisfaction with Sexual Assault Prevention and Response Training ... 48
Conclusion..49

CHAPTER SIX
Branch of Service Differences on Measures of Sexual Assault and Sexual
   Harassment 51

CHAPTER SEVEN
Findings from the Coast Guard Reserve 59
Sexual Assault 59
Sexual Harassment and Gender Discrimination 61

CHAPTER EIGHT
Discussion and Recommendations...63
Sexual Assault 63
Sexual Harassment and Gender Discrimination 64
Recommendations 67
Additional Information on the RAND Military Workplace Study............................68

APPENDIX
The Coast Guard Sample..69

References..79

Figures and Tables

Figures

4.1. Percentage of Active-Component Coast Guard Members Who Experienced Sexual Harassment in the Past Year, by Gender and Pay Grade 29

4.2. Percentage of Active-Component Coast Guard Members Who Experienced Gender Discrimination in the Past Year, by Gender and Pay Grade 30

4.3. Proportion of Service Members Experiencing Sexual Harassment and Gender Discrimination and the Relative Overlap Between These Military Equal Opportunity Violations 31

4.4. Percentage of Active-Component Coast Guard Women and Men Who Experienced Each Type of Sexual Harassment and Gender Discrimination Violation in the Past Year 36

Tables

2.1. Coast Guard Active-Component Sample 5

3.1. Estimated Percentage of Active-Component DoD and Coast Guard Service Members Who Experienced Any Type of Sexual Assault in the Past Year, by Gender and Service 10

3.2. Estimated Percentage of Active-Component Coast Guard Service Members Who Experienced a Sexual Assault in the Past Year, by Gender and Type of Assault 11

3.3. Estimated Percentage of Active-Component DoD and Coast Guard Service Members Who Experienced Any Type of Unwanted Event, by Gender and Service Branch 13

3.4. Estimated Percentage of Active-Component DoD and Coast Guard Service Members Who Experienced a Sexual Assault in Their Lifetime, by Gender and Service Branch 16

3.5. Estimated Percentage of Active-Component DoD and Coast Guard Service Members Who Experienced a Sexual Assault Prior to Joining the Military, by Gender and Service 16

3.6. Estimated Percentage of Active-Component DoD and Coast Guard Service Members Who Experienced a Sexual Assault Since Joining the Military, by Gender and Service Branch 17

4.1.   Estimated Percentage of Active-Component Coast Guard and DoD Service Members Who Experienced a Sexually Hostile Work Environment in the Past Year, by Gender and Service Branch 24

4.2.   Estimated Percentage of Active-Component Coast Guard and DoD Service Members Who Experienced Sexual *Quid Pro Quo* in the Past Year, by Gender 25

4.3.   Estimated Percentage of Active-Component Coast Guard and DoD Service Members Who Experienced Sexual Harassment in the Past Year, by Gender and Service Branch 26

4.4.   Estimated Percentage of Active-Component Coast Guard and DoD Service Members Who Experienced Gender Discrimination in the Past Year, by Gender and Service Branch 27

4.5.   Estimated Percentage of Active-Component Coast Guard and DoD Service Members Who Experienced Sexual Harassment or Gender Discrimination in the Past Year, by Gender and Service Branch 28

4.6.   Estimated Percentage of Active-Component Coast Guard Members Who Experienced Each Type of Inappropriate Workplace Behavior in the Past Year, by Gender 32

4.7.   Estimated Percentage of Active-Component Coast Guard Members Who Experienced Each Type of Sexual Harassment (Hostile Workplace or *Quid Pro Quo*) or Gender Discrimination Violation in the Past Year 34

4.8.   Characteristics of the Situation and Offenders 38

4.9.   Self-Reported Likelihood of Choosing to Stay on Active Duty Among Coast Guard Members Who Had Experienced Either Sexual Harassment or Gender Discrimination in the Past Year 40

4.10.   Satisfaction with Response to Report of Sexual Harassment or Gender Discrimination 42

4.11.   Barriers to Reporting Sexual Harassment and Gender Discrimination 43

5.1.   Perceptions of Safety at Home Duty Station, Estimated Percentages by Gender 45

5.2.   Perceptions of Frequency of Sexual Harassment in the Military, Estimated Percentages by Gender 46

5.3.   Perceptions of Frequency of Sexual Harassment in the Military, Estimated Percentages by Service 46

6.1.   Factors Considered as Possibly Explaining Service Differences in the Rate of Sexual Assault and Sexual Harassment 53

6.2.   Adjusted and Unadjusted Risk Ratios for Sexual Assault Relative to Coast Guard Personnel, by Gender and Service 54

6.3.   Adjusted and Unadjusted Risk Ratios for Sexual Harassment Relative to Coast Guard Personnel, by Gender and Service 57

7.1.   Estimated Percentage of Coast Guard Reserve Members Who Experienced a Sexual Assault in the Past Year, by Gender and Assault Type 60

7.2.   Estimated Percentage of Coast Guard Reserve Members Who Experienced Any Type of Sexual Assault in the Past Year, by Gender and Pay Grade 60

7.3.   Estimated Percentage of Coast Guard Reserve Members Who Experienced a Sex-Based MEO Violation in the Past Year, by Gender and Type 61

A.1.   Coast Guard Active-Component Sampling Frame and Sample Sizes, by Gender, Service, and Pay Grade 70

A.2.   Case Disposition Frequencies for the Coast Guard Active-Component Sample 71

A.3.   Quality of Mailing Address Based on Initial Mailing 72

A.4.   Quality of Email Address Based on Initial Email 72

A.5.   Case Disposition Frequencies for Coast Guard Reserve Sample 73

A.6.   Response Rates by Form Type for the Coast Guard Active Component 74

A.7.   Response Rates for the Coast Guard Active Component, by Gender and Pay Grade 74

A.8.   Response Rates for the Coast Guard Reserve, by Form 75

A.9.   Response Rates for the Coast Guard Reserve, by Gender and Pay Grade 75

A.10. Balance of Weighted Respondents to the Coast Guard Active-Component Population 76

Summary

In early 2014, the Department of Defense (DoD) asked the RAND National Defense Research Institute to conduct an independent assessment of sexual assault, sexual harassment, and gender discrimination in the military—an assessment last conducted in 2012 by the department itself through the Workplace and Gender Relations Survey of Active Duty Members. Shortly thereafter, the Coast Guard requested that RAND expand the study to include an assessment of its active and reserve forces as well.

This report provides estimates for the Coast Guard active and reserve components from the resulting study, the RAND Military Workplace Study (RMWS), which invited close to 14,000 active-component Coast Guard members and all 7,592 Coast Guard Reserve members to participate in a survey fielded in August and September of 2014. High rates of participation by sampled Coast Guard members resulted in more than 7,000 survey responses from active-component members, including more than one-half of all active-component women. We also received approximately 2,500 survey responses from Coast Guard Reserve members. Because the survey was also conducted with active- and reserve-component members of the Army, Navy, Air Force, and Marine Corps, in many cases we are able to compare Coast Guard findings with those from the DoD services.

Compared to prior DoD studies, the RMWS took a new approach to counting individuals in the military who experienced sexual assault, sexual harassment, or gender discrimination. Our measurement of sexual assault aligns closely with the definitions and criteria in the Uniform Code of Military Justice (UCMJ) for Article 120 and Article 80 crimes.[1] The survey measures of sexual harassment and gender discrimination use criteria drawn directly from DoD Directive 1350.2 on military equal opportunity (MEO) violations. Compared with past surveys that were designed to measure a climate of sexual misconduct associated with illegal behavior, our approach offers greater precision in estimating the number of *crimes* and *MEO violations* that have occurred. Specifically, the RMWS measures

---

[1] Article 120 of the UCMJ, "Rape and Sexual Assault Generally," defines four offenses: rape, sexual assault, aggravated sexual contact, and abusive sexual contact. In this report, as in the title of Article 120, we use the term *sexual assault* to refer to all four offenses, not just to the one offense labeled sexual assault.

- *sexual assault*, which captures three mutually exclusive categories: *penetrative*, *non-penetrative*, and *attempted penetrative* crimes
- *sexual harassment*, which consists of
  - *sexually hostile work environment*—a workplace characterized by severe or pervasive unwelcome sexual advances, comments, or physical conduct that offends service members
  - *sexual quid pro quo*—incidents in which someone uses his or her power or influence within the military to attempt to coerce sexual behavior in exchange for a workplace benefit
- *gender discrimination*—incidents in which service members are subject to mistreatment on the basis of their gender that affects their employment conditions.

As with all crime victim surveys, we classify service members as experiencing sexual assault, sexual harassment, or gender discrimination based on their memories of the event as expressed in their survey responses. It is likely that a full review of all evidence would reveal that some respondents whom we classify as not having experienced a sexual assault, sexual harassment, or gender discrimination based on their survey responses actually did have one of these experiences. Similarly, some whom we classify as having experienced a crime or violation may have experienced an event that would not meet the minimum DoD criteria. A principal focus of our survey development was to minimize both of these types of errors, but they cannot be completely eliminated in a self-report survey.

Sexual Assault: Active Component

We estimate that between 180 and 390 of the more than 39,000 active-component Coast Guard members experienced a criminal sexual assault in the past year. This represents approximately 0.7 percent of Coast Guard members, including 3 percent of women and 0.3 percent of men. Because some members experienced multiple incidents, the past-year incidence rates are necessarily higher than these past-year prevalence rates. Specifically, while 0.7 per 100 Coast Guard members experienced one or more sexual assaults in the past year, there were approximately 1.7 separate incidents in the past year per 100 Coast Guard members. The prevalence rates are low compared with those of the Army, Navy, and Marine Corps, but similar to those of the Air Force. Indeed, even after accounting for demographic and other differences between members of each service, women in the Army, Navy, and Marine Corps are more than twice as likely to have been sexually assaulted in the past year, and men in those services are four to five times as likely to have been sexually assaulted in comparison to women and men in the Coast Guard.

Because of the comparatively low rate of sexual assault for Coast Guard men, there are too few with past-year sexual assaults for us to characterize their experiences in detail. Therefore, we limit our discussion to the experiences of women with sexual assaults in the past year.

When women in the Coast Guard were assaulted in the past year, the assailant was another member of the military in 77 percent of all cases. This rate is significantly lower than the proportion of women assaulted by a member of the military across all DoD services (89 percent), although the proportion among sexually assaulted women in the Coast Guard is similar to the Air Force. We are not able to estimate the proportion of Coast Guard members who experienced retaliation after officially reporting a sexual assault in the last year. This is because of the low numbers of respondents who officially reported a sexual assault.

When a sexual assault occurs against Coast Guard women, alcohol is more frequently involved than among women in most other DoD services. Indeed, more than 75 percent of assaults against Coast Guard women occurred after either the woman or the assailant had been drinking, and usually both had been. In contrast, 56 percent of assaults against women in DoD services occurred after alcohol consumption by the woman or the assailant. This higher proportion of sexual assaults involving alcohol is consistent with other results showing that Coast Guard women are at lower risk of sexual assault at work than women in some other services. For example, assaults against Coast Guard women more commonly occur while out with friends or at a party.

Sexual Harassment and Gender Discrimination: Active Component

Far more Coast Guard members experienced sexual harassment or gender discrimination in the past year than experienced a sexual assault. We estimate that approximately 6 percent of active-component Coast Guard members, or 2,350 members, experienced sexual harassment in the past year. A higher proportion of women (1 out of 5) than men (1 out of 25) had workplace experiences in the past year that under Coast Guard directives would be classified as sexual harassment.

That sexual harassment is relatively common within the Coast Guard is widely understood by its members, at least by its female members. Specifically, across active-component members, 71 percent of women and 39 percent of men indicated that sexual harassment in the military is either "common" or "very common." These rates are comparable to those found across DoD services, where 76 percent of women and 45 percent of men describe sexual harassment as "common" or "very common."

Although less common than sexual harassment, approximately 1,020 active-component members of the Coast Guard experienced gender discrimination, with women 11 times more likely than men to be classified as having such an experience in the past year. Like sexual harassment, gender discrimination against women is widely recognized as an issue for the Coast Guard, at least among women, 62 percent of whom describe such discrimination as "common" or "very common" in the military, compared with 27 percent of men.

The substantial majority of Coast Guard members who experienced sexual harassment or gender discrimination described their offender(s) as members of the military (90 percent). In two-thirds of the incidents involving a military service member, one or more of the offenders were of higher rank than the target, and the offender(s) in more than one-half of incidents was reportedly a supervisor or unit leader.

Sexual harassment and gender discrimination may also contribute to the risk of sexual assault. Certainly the correlation between the two is strong, as those women who experienced sexual harassment in the past year were far more likely than those who were not sexually harassed to have also experienced a sexual assault during the same period. Moreover, 30 percent of women who were assaulted indicated that their assailant previously sexually harassed them.

Experiences of the Coast Guard Reserve

Sexual assault is less common in the Coast Guard Reserve. We estimate that approximately 40 individuals were assaulted on or off duty in the past year, or just under one-half of one percent of the more than 7,500 members of the Coast Guard Reserve who are below the rank of flag officer. Rates of past-year sexual assault for Coast Guard men and women in the reserves are not significantly different than rates found for DoD reserve-component members. The majority of all Coast Guard Reserve members who experienced a sexual assault were women.

Rates of MEO violations in the past year are significantly lower for members of the Coast Guard Reserve than for the active component. We estimate that approximately 4 percent of reservists experienced sexual harassment or gender discrimination in the past year. The risk for such violations varied substantially by gender, with 2 percent of men and 15 percent of women experiencing these violations, most of which involved a sexually hostile workplace environment.

Recommendations

Based on the results of our survey analyses, we offer the following recommendations.

1. *Concentrate additional prevention and enforcement on sexual harassment and gender discrimination.* Reducing the incidence of sexual harassment and gender discrimination is likely to have far-reaching benefits for the Coast Guard, possibly including improved workplace productivity, reduced sick time, and improved recruitment and retention, and it may reduce the prevalence of sexual assault.

2. *Review training and instructional materials to ensure that they make clear that some reportable sexual assaults may occur in the context of hazing or bullying, and so may not be perceived by either the service member or the offender as a sexual encounter.* Ensuring that members of the Coast Guard understand the full scope of physical assaults that qualify as sexual assaults may improve reporting and provide those who are being abused with needed response systems.

3. *Develop monitoring systems for sexual harassment, gender discrimination, hazing, bullying, and physical assaults.* The prevalence of sexual assault in the Coast Guard is sufficiently high that it is possible to estimate the extent of the problem from smaller numbers of individuals—including, for instance, members assigned to individual commands, installations, or possibly ships. We believe it might be valuable to extend this monitoring to cover not only MEO violations, but also hazing, bullying, and physical assaults, all of which form a nexus that may contribute to sexual assault risk and to undermining good order and discipline in the Coast Guard.

4. *Investigate the causes and consequences of sexual assault.* The RMWS has provided unprecedented detail on the nature and circumstances of sexual assault, sexual harassment, and gender discrimination in the military services, but the new insights offered by these data raise new questions that we believe the Coast Guard should consider investigating further:

   a. We find significant differences between the risk of sexual assault to which Coast Guard members are exposed and that for members of the Army, Navy, and Marine Corps. Although we have ruled out many plausible risk factors on which members of each of these services may differ from the Coast Guard, we have not identified what does explain the Coast Guard's lower risk. If we were able to determine that risk differences are attributable to cultural differences between the services, differences in training, differences in patterns of life members experience (such as where they are quartered or the amount of time they spend away from home), or other such factors, this could provide important insights into how to further reduce sexual assault risk in the Coast Guard, in other military services, and possibly in civilian settings as well.

   b. Our results raise the possibility that sexual harassment and gender discrimination may have a range of harmful effects on service members' careers, their safety, and their retention in the Coast Guard. A longitudinal study of service members' responses to sexual harassment and discrimination would be a helpful adjunct to these data to better estimate the consequences for the Coast Guard of these events.

Acknowledgments

We wish to thank the servicemen and -women who took the time to complete the RAND Military Workplace Study survey and share their experiences, even when those experiences were painful to recount.

Many people assisted us with the development of the new survey instrument. The leadership and staff in the U.S. Coast Guard's and each of the services' sexual assault prevention and response offices provided many rounds of review, valuable suggestions, and feedback, as did research staff from the Air Force and Army Research Institutes. James Clark and Teresa Scalzo consulted on the interpretation of Article 120 of the Uniform Code of Military Justice. Jonathan Welch, of RAND, and Tom Bush consulted on the National Guard and reserve form of the survey. The questionnaire further benefited from comments of 24 anonymous service members and recent veterans who agreed to pre-test the survey and provide us with their reactions to it. We express our appreciation to John Boyle, Senior Vice President of ICF International; Richard Baskin, from Decipher Inc.; and Mary Koss, from the University of Arizona, for providing a survey instrument they developed to measure sexual assault experiences among service members in the U.S. Air Force. This instrument was helpful to us in creating the new survey instrument used in this study.

In addition to assisting us with the development of the survey instrument, the members of our scientific advisory board have provided invaluable guidance on difficult decisions throughout the project.

We are grateful for the assistance and expert advice provided to us by the Defense Manpower Data Center (DMDC), and especially to Elizabeth Van Winkle, who shared DMDC experience from prior administrations of the Workplace and Gender Relations surveys, and who served as a liaison between RAND and other parts of DMDC. We also thank Paul Rosenfeld for his rapid and careful reviews of the survey licensing materials submitted by RAND to the Office of the Secretary of Defense; Major Brandi Ritter, in the Office of the Under Secretary of Defense for Personnel and Readiness Research Regulatory Oversight Office; and Carlos Comperatore, Chair, Coast Guard Institutional Review Board, for their review and oversight of study human subjects protections.

We have benefited from a strong and critical set of internal and external quality assurance reviewers, including Cynthia Cook, Greg Ridgeway, John Winkler, Bernie Rostker, and Daniel Ginsberg, all of whom have provided valuable guidance throughout this effort. We also thank Lane Burgette for assistance with double-checking and troubleshooting our statistical programming.

We also wish to thank the interviewers and helpdesk staff at Westat, who supported the survey through its fielding period.

Finally, we wish to thank our U.S. Coast Guard and DoD sponsors in the Sexual Assault Prevention and Response Offices. Captain Benjamin L. Smith, Nate Galbreath, and Major General Jeffrey Snow provided strong support to the study team throughout the project.

Abbreviations

AFMS	active federal military service
AFQT	Armed Forces Qualifying Test
CI	confidence interval
DMDC	Defense Manpower Data Center
DoD	Department of Defense
MEO	military equal opportunity
NCOA	National Change of Address
NR	not reportable
RMWS	RAND Military Workplace Study
RR1	American Association for Public Opinion Research response rate 1
SE	standard error
TAD	temporary additional duty
TDY	temporary duty
UCMJ	Uniform Code of Military Justice
WRGA	Workplace and Gender Relations Survey of Active Duty Members

CHAPTER ONE

Introduction

Andrew R. Morral, Kristie L. Gore, and Terry L. Schell

In early 2014, the Department of Defense (DoD) asked the RAND National Defense Research Institute to conduct an independent assessment of sexual assault, sexual harassment, and gender discrimination in the military—an assessment last conducted in 2012 by the department itself through the Workplace and Gender Relations Survey of Active Duty Members (WGRA). The 2014 RAND Military Workplace Study (RMWS) is based on a much larger sample of the military community than in previous surveys—men and women, active and reserve components, and including the four DoD military services plus the Coast Guard—and is designed to more-precisely estimate the total number of service members experiencing sexual assault, sexual harassment, and gender discrimination.

The objectives of the 2014 survey were to

- establish precise and objective estimates of the percentage of service members who experience sexual assault, sexual harassment, and gender discrimination
- describe the characteristics of these incidents, such as where and when they occurred, who harassed or assaulted the member, whether the event was reported, and what services the member sought
- identify barriers to reporting these incidents and barriers to the receipt of support and legal services.

On December 5, 2014, RAND released preliminary *top-line* results from this survey. These top-line numbers referred to the broadest categories of outcomes and included only estimated numbers and percentages of active-component Coast Guard members who experienced sexual assault, sexual harassment, and gender discrimination in the past year by gender, service, and type of offense. This report expands on the findings presented in the top-line report to include information on

- the samples, response rates, and survey weights
- top-line and detailed results for Coast Guard Reserve members
- the context and perpetrators of sexual assault and harassment

- factors that explain some of the service differences observed in rates of sexual assault
- recommendations for better understanding and prevention of sexual assault and harassment in the Coast Guard.

In this third volume of the series on *Sexual Assault and Sexual Harassment in the U.S. Military*, we present these findings and analyses for the U.S. Coast Guard and Coast Guard Reserve. Volume 2 provides detailed results for Army, Navy, Air Force, and Marine Corps active and reserve components. Volume 4 will provide analyses designed to evaluate the likely effects of survey nonresponse or other types of biases on our population estimates. Annexes to Volumes 2 and 3 contain detailed tabular results for the DoD active component and for the Coast Guard active component, respectively.

Chapter Two begins with an overview of the study design and analysis approach. We then present key findings from our analyses of sexual assault in the Coast Guard (Chapter Three) and sexual harassment and gender discrimination in the Coast Guard (Chapter Four). Chapter Five describes Coast Guard members' beliefs and attitudes about sexual assault and sexual harassment. Chapter Six investigates possible explanations for the observed differences among the service branches on rates of sexual assault and sexual harassment. Chapter Seven presents sexual assault and harassment findings from the reserve component, including comparisons between the active and reserve components. The final chapter draws broader conclusions across the individual chapters and presents recommendations for consideration. In addition, the appendix contains more details of the study design, describing the characteristics of the sampled service members and their representativeness of the overall military population. An annex to this volume contains detailed data on Coast Guard members' experiences of sexual assault and military equal opportunity (MEO) violations, and on beliefs about sexual assault and sexual harassment prevalence, prevention, and progress.

CHAPTER TWO

Study Design and Analysis Approach

Terry L. Schell and Bonnie Ghosh-Dastidar

Volume 1 of this series, *Sexual Assault and Sexual Harassment in the U.S. Military: Design of the 2014 RAND Military Workplace Study,* was released in December of 2014, along with the top-line results. Volume 1 details the context and many of the methods we used for the RMWS, including discussions of the challenges associated with measuring sexual assault and sexual harassment, the strategies we used to improve the precision with which we estimated these phenomena, the development of the survey questionnaire, the survey sampling design, and the weighting methods. Volume 1 also contains the survey questionnaires used. In this chapter, we provide an overview of our survey design and sample, survey response rates, and the statistical analysis and reporting conventions used in this report (Volume 3). The appendix contains additional details on the Coast Guard sample and response rates. For a more-detailed discussion of survey methodology, we refer readers to Volume 1. For additional information about potential sources of bias in the estimates, we refer the reader to Volume 4, which includes results from studies of survey nonresponders.

Study Design and Sample

DoD, in consultation with the White House National Security Staff, stipulated that the sample size for the RMWS was to include a census of all women and 25 percent of men in the active component of the Army, Navy, Air Force, and Marine Corps. In addition, we were asked to include a smaller sample of National Guard and reserve members sufficient to support comparisons of sexual assault, sexual harassment, and gender discrimination between the active and reserve components. Subsequently, the U.S. Coast Guard also asked that RAND include a sample of its active- and reserve-component members. In total, therefore, RAND invited close to 560,000 service members to participate in the study, making it the largest study of sexual assault and harassment ever conducted in the military.

The large sample for this study is particularly valuable for understanding the experiences of relatively small subgroups in the population. For example, RAND's survey provides more information about the experiences of DoD men who have been sexually assaulted than prior studies. The large sample also gave RAND the opportunity to test how changing the questionnaire itself affects survey results. Specifically, we were able to use a segment of our overall sample to draw direct comparisons between rates of sexual assault and sexual harassment as measured using the 2014 RMWS questionnaire and the measures used in the 2012 WGRA questionnaire.

To enable this comparison and others, we randomly assigned respondents to one of three different survey questionnaires. The size of the Coast Guard and our sample of its members were not large enough to support precise estimates on both the WGRA and RMWS measures, so all Coast Guard members were randomly assigned to one of the new RMWS questionnaires.

1. A "long form," consisting of a sexual assault module; a sex-based MEO violation module, which assessed sexual harassment and gender discrimination; and questions on respondent demographics, psychological state, command climate, attitudes and beliefs about sexual assault in the military and the nation, and other related issues.
2. A "medium form," consisting of the sexual assault module, the sex-based MEO violation module, and demographic questions.
3. A "short form," consisting of the sexual assault module, the screening items from the sex-based MEO violation module, and demographic questions. Thus, these respondents did not complete the full, sex-based MEO violation assessment.

Multiple versions of the RAND form (long, medium, and short versions) were used to minimize respondent burden and costs to the Coast Guard. It was not necessary to collect general experiences and attitudes from the entire sample to derive precise results, and doing so would have been wasteful of service members' time. Therefore, we designed the survey so that each question was posed to only as many service members as was necessary to provide the precision required for the question. In general, those items that concern relatively rare events (such as sexual assault in the past year) must be asked of the largest number of people to arrive at precise estimates, whereas questions concerning attitudes or beliefs, for instance, which everyone can answer, need only be asked of a comparatively small sample. Similar to the DoD reserve-component samples discussed in Volume 2, the relatively small Coast Guard Reserve sample was always assigned to either the medium or short forms of the RMWS questionnaire.

Active-Component Sample and Response Rates

A total of 14,167 members of the Coast Guard active component were randomly selected from a population of 39,112 Coast Guard members who were not members of the Coast Guard Reserve and who met the study inclusion criteria requiring that they be age 18 or older, below the rank of a flag officer, and in service for at least six months

Table 2.1
Coast Guard Active-Component Sample

	Total		Women		Men	
	Population	Sample	Population	Sample	Population	Sample
Total	39,112	14,167	5,852	5,852	33,260	8,315
Pay grade						
E1–E4	12,158	4,937	2,515	2,515	9,643	2,422
E5–E9	20,345	6,625	2,047	2,047	18,298	4,578
O1–O3	3,859	1,638	900	900	2,959	738
O4–O6	2,750	967	390	390	2,360	577

NOTE: Sample contains both respondents and nonrespondents. Population refers to the study eligible population.

(Table 2.1). This follows the procedures used in prior WGRA surveys. The sample included 5,852 women and 8,315 men.

A total of 7,307 active-component Coast Guard members completed the RMWS survey, or just over 51 percent of the sample. This is substantially higher than the DoD response rate of 30 percent. The respondents included 53 percent of the women sampled (3,106) and 51 percent of the men (4,201). Across pay grades, senior officers (O4–O6) had a response rate (71 percent) considerably higher than that of junior enlisted (E1–E4), who had the lowest response rate (43 percent).

Reserve Component Sample and Response Rates

Due to the small size of the Coast Guard Reserve population, we included every eligible reserve member in the survey. The same eligibility criteria used in the active component (described previously) was also used in the Coast Guard Reserve. The Coast Guard Reserve sample (and sample frame) totaled 7,592 members, including 1,267 women and 6,325 men.

The response rate for the Coast Guard Reserve sample was 33.4 percent, almost 20 percentage points lower than the 51.6 percent response rate for the active component. However, this response rate is higher than the DoD reserve-component response rate (22.6 percent). The response rate for women in the Coast Guard Reserve (38.0 percent) was higher than that for men (32.4 percent).

Statistical Analysis and Reporting Conventions Used in This Report

The statistical analyses presented in this report, its appendix, and the Annex to Volume 3 employ statistical procedures designed to reduce the likelihood of drawing inappropriate conclusions or compromising the privacy of respondents.

First, we assured respondents in the survey *Privacy Statement* (part of the informed consent) that our reports would not include analyses conducted with subsets smaller than 15 respondents. To maintain participant privacy, the report and its annex do not include sample statistics (including confidence intervals) computed for groups smaller than 15 unweighted respondents. If such a cell appears in a table, the point estimates and its confidence intervals are replaced with NR, or "not reportable."

Second, the report contains estimated population percentages that vary dramatically in their statistical precision. Some estimates have a 95-percent confidence interval that have a width of 0.3 percentage points, while others have a width of 30 percentage points. This occurs because some percentages are estimated using more than 100,000 respondents, while others are estimated on small subsamples (e.g., male airmen who experienced a sexual assault). To reduce the likelihood of misinterpretations, percentages with very low precision are not reported. Specifically, percentages estimated with a margin of error greater than 15 percentage points are replaced with NR (where the margin of error is defined as the larger half-width of the confidence interval). In such cases, the confidence intervals are still presented to communicate the range of percentages that are consistent with the data. Such imprecise estimates are better thought about as ranges rather than points.

The text and tables in this report do not use a constant level of numerical precision. Because the statistical precision of the estimates vary by over two orders of magnitude, and because the purpose of numbers presented in the text and in tables may be slightly different, we have tried to select a level of numerical precision that is appropriate for each situation. In contrast to the variation in numerical precision within the body of the report, the annex presents percentages to two decimal places. The reader is cautioned to interpret these estimates with respect to their confidence intervals rather than their apparent numerical precision. In general, the report includes confidence intervals (either in the body of the report or in the annex) for all of the statistics that are interpreted as population estimates.

To streamline presentation, the report focuses primarily on large effects or large differences between groups. With large differences, formal tests of statistical significance are not included in the text, because significance can be inferred from non-overlapping confidence intervals. In some cases, we do include *p*-values in the text or use indicators of statistical significance in tables. This is done when we explicitly tested a hypothesis that cannot be investigated directly with the confidence intervals presented (e.g., comparing one service to the average of the other three), or when the confidence intervals overlap but the differences are still statistically significant. Whenever a difference between estimates is discussed in the text it is statistically significant, unless explicitly noted to be not statistically significant. In general, claims about statistical significance in the text refer to a standard $\alpha = 0.05$, two-tailed test. In some analyses involving variables with more than two levels, Bonferroni corrections for multiple testing have been used. When used, the Bonferroni correction is noted in the text or table.

All estimates presented in the report and its annex (unless specifically labeled otherwise), use survey weights that account for the sample design and survey non-response. As discussed in Volume 1, estimates derived from measures used in prior WGRA surveys are analyzed using weights that were derived similar to those used in prior WGRA studies. All other analyses used the RAND-designed survey weights outlined in Volume 1. Volume 4 provides additional information about, and analyses of, these weights.

Confidence intervals for proportions are computed as exact binomials (Clopper-Pearson). Confidence intervals for counts or continuous values are computed using the standard normal approximation. Variance estimation is typically done with the Taylor series linearization method. However, that method cannot be used to estimate the variance of a percentage with a zero numerator. In those cases, confidence intervals were computed using the Hanley and Lippman-Hand (1983) method with the sample size defined using the Kish (1965) estimate for effective sample size.

CHAPTER THREE
Sexual Assault Findings: Coast Guard Active Component

Lisa H. Jaycox, Terry L. Schell, Andrew R. Morral, Amy Street,
Coreen Farris, Dean Kilpatrick, and Terri Tanielian

The RMWS survey contains a detailed assessment of sexual assault designed to correspond to the legal criteria specified in Article 120 of the Uniform Code of Military Justice (UCMJ). To be classified as having experienced a sexual assault, respondents must first have indicated that they experienced one of six anatomically specific unwanted behavioral events. If they indicated that one of these events occurred in the past year, they were then asked a series of additional questions designed to assess (a) whether the event was intended either for a sexual purpose, to abuse, or to humiliate, as indicated in the UCMJ, and (b) whether the offender used one of the coercion methods specified in the UCMJ as defining a criminal sex act. The complete survey instrument and a detailed discussion of the rationale behind this approach to assessing sexual assault may be found in Volume 1 of this series.

Sexual Assault Prevalence

The RMWS estimates suggest that 0.69 percent of the active-component Coast Guard population experienced at least one sexual assault in the past year (Table 3.1). We estimate that the total number of Coast Guard members in our sample frame who experienced a sexual assault in the past year is about 270 (95% CI: 180–390).[1] The sample frame consisted of all active-component Coast Guard members who (as of May 1, 2014) were at least 18 years of age, had served six months or more, and were below the pay grade of a flag officer. The estimated rate of sexual assault varied by gender: Approximately 3 in 1,000 men and 30 in 1,000 women were sexually assaulted in the past year. Because of this difference in risk, the majority of those who were sexually assaulted were women, even though women represent a minority of the overall Coast Guard population. We estimate there were 170 women (95% CI: 130–220) and 100

---

[1] The confidence interval (CI) describes the range within which the true value for the population is likely to lie, based on the data available in the sample. In the case of a 95 percent CI, we expect that the true population value is within the given range 95 percent of the time.

men (95% CI: 30–240) in the Coast Guard who experienced a sexual assault in the past year. As seen in Table 3.1, these rates are significantly lower than our estimates for the percentage of active-component Army, Navy, and Marine Corps members who experienced a sexual assault in the past year.[2]

Breakdowns of the number of assaults within the Coast Guard by pay grade can be seen in the Annex to Volume 3, Table A.1, showing no overall distinction between pay grades partly due to wide confidence intervals in this small sample. Among DoD servicemen and -women (see Volume 2), junior enlisted members (E1–E4) had a higher risk of sexual assault in the past year than senior enlisted members (E5–E9, W1–W5). Among officers, junior and senior DoD servicemen had comparable rates of sexual assault in the past year, but junior-grade DoD servicewomen had more than twice the rate of sexual assaults in the past year as did more-senior-grade women.

To gain a better understanding of the nature of these events, we broke down the overall rate of sexual assault into the specific type of sexual assault that the respondent was classified as experiencing (Table 3.2). Although all respondents answered all six sexual assault screener items, the survey instrument was structured so that if a respondent was classified as having experienced a penetrative sexual assault, they skipped the detailed subsequent questions about lesser offenses. Similarly, if they qualified as having experienced a non-penetrative sexual assault, they skipped the final follow-up

Table 3.1
Estimated Percentage of Active-Component DoD and Coast Guard Service Members Who Experienced Any Type of Sexual Assault in the Past Year, by Gender and Service

Service	Total	Men	Women
Coast Guard	0.69% (0.46–1.00)	0.29% (0.09–0.71)	2.97% (2.25–3.83)
Army	1.46%[a] (1.25–1.70)	0.95%[a] (0.72–1.23)	4.69%[a] (4.30–5.09)
Navy	2.36%[a] (1.92–2.86)	1.48%[a] (1.00–2.12)	6.48%[a] (5.79–7.22)
Air Force	0.78% (0.70–0.87)	0.29% (0.21–0.39)	2.90% (2.67–3.15)
Marine Corps	1.63%[a] (1.15–2.24)	1.13%[a] (0.65–1.84)	7.86%[a] (6.65–9.21)

NOTE: 95-percent confidence intervals for each estimate are indicated in parentheses.
[a] Percentage is significantly different from Coast Guard within a column; $p < 0.05$, Bonferroni corrected.

---

[2] For each of these comparisons, we use a $p < 0.05$, Bonferroni corrected, for four comparisons (Coast Guard to each DoD service).

Table 3.2
Estimated Percentage of Active-Component Coast Guard Service Members Who Experienced a Sexual Assault in the Past Year, by Gender and Type of Assault

Service	Total	Men	Women
Any sexual assault	0.69% (0.46–1.00)	0.29% (0.09–0.71)	2.97% (2.25–3.83)
Penetrative sexual assault	0.36% (0.18–0.65)	0.17% (0.02–0.60)	1.44% (0.93–2.12)
Non-penetrative sexual assault	0.33% (0.20–0.50)	0.12% (0.02–0.35)	1.50% (1.03–2.12)
Attempted penetrative	0.00% (0.00–0.06)	0.00% (0.00–0.20)	0.03% (0.00–0.17)

NOTES: There were no cases of attempted penetrative assault among men in the sample. 95-percent confidence intervals for each estimate are indicated in parentheses.

questions assessing whether they experienced an attempted penetrative sexual assault. Thus, the instrument defines three mutually exclusive categories of sexual assault: *penetrative, non-penetrative,* and *attempted penetrative.*[3]

Penetrative sexual assaults are events that people often refer to as rape, including penetration of the mouth, anus, or vagina by a penis, body part, or object. We describe the measure as *penetrative sexual assault* in order to include both penetrative assaults that would be charged as rape and penetrative assaults that would be charged as sexual assault. *Non-penetrative assaults* include incidents in which private areas on the service member's body are touched without penetration, or where the service member is made to have contact with the private areas of another person's body.[4] The *attempted penetrative sexual assault* category applies only to those people who could not be classified as experiencing crimes that could be charged directly via UCMJ Article 120 (i.e., *penetrative* or *non-penetrative sexual assaults*). That is, they indicated having experienced an event in which someone attempted to sexually assault them (charged via UCMJ Article 80), but the person never made physical contact with a private area of their body (which would have allowed categorization under the *non-penetrative sexual assault* category). This approach to classifying sexual assaults results in nearly all sexual assaults being categorized as either *penetrative* or *non-penetrative*, with very few classified as *attempted* assaults. A detailed analysis of how individuals answered the sexual assault screening items, and thus were classified as having experienced a sexual assault, can be found in Volume 4 of this series.

---

[3] An implication of this strategy is that once a service member indicated having experienced a sexual assault during the past year, we did not continue to ask detailed questions that would have identified additional sexual assaults. A detailed analysis of the sexual assault instrument, including its correspondence with the specific wording of Article 120 of the UCMJ, is included in Volume 1 of this series.

[4] *Private areas* were defined to include the buttocks, inner thigh, breast, groin, anus, vagina, penis, and testicles.

Within the DoD sample, the distribution across type of assault was similar for men and women, with approximately one-half of all sexual assaults being classified as *penetrative sexual assaults*. This is a higher estimated rate of penetrative assaults than in 2010, when approximately 25 percent of all assaults against active-component women and 21 percent of assaults against active-component men were classified as penetrative. This difference likely resulted from differences between the RMWS and WGRA measurement approaches, rather than from changes in the true prevalence of penetrative sexual assaults. Our analyses of the results of the DoD experiment in which some members received the old WGRA questions and some received the new RMWS questions suggest that the new questions identify more penetrative crimes than the old questions.

Among individuals who experienced at least one sexual assault in the past year, about one-half indicated it was a single event, with an overall mean of 2.44 incidents (95% CI: 1.66–3.23) in the past year across both men and women.[5] In the DoD sample (see Volume 2), we observed that sexually assaulted men reported more incidents in the past year, on average, than sexually assaulted women. However, within the Coast Guard sample there were too few sexually assaulted men to analyze the number of incidents in the past year by gender. Because some members experienced multiple incidents, the past-year incidence rates are necessarily higher than the past-year prevalence rates provided in Table 3.2. Specifically, while 0.69 per 100 Coast Guard members experienced one or more sexual assaults in the past year, there were 1.68 (95% CI: 0.64–2.71) separate incidents in the past year per 100 Coast Guard members. Among Coast Guard women, we estimate that there were 5.23 incidents per 100 members (95% CI: 3.66–6.79).

---

[5]   The variable used to estimate the average number of sexual assaults experienced in the past year (SAFU1) included six response options. Four of the responses are numeric responses (1–4 times), but two responses are non-numeric: "5 or more times since [X date]" and "More than once, but not sure the number of times it happened since [X date]." To calculate the mean number of sexual assaults, we used the most conservative approach to coding the non-numeric responses. Respondents who indicated that they experienced a sexual assault "5 or more times since [X date]" were coded as experiencing five incidents. Respondents who indicated that they experienced sexual assault "More than once, but not sure the number of times it happened since [X date]" were coded as experiencing two incidents. Thus the number of incidents is computed in a conservative manner that will undercount incidents for those individuals who had more than 5 in the past year. However, it is also important to note that some of the incidents we are counting may not qualify as sexual assault crimes under the UCMJ. The survey established that at least one incident per respondent qualified as a crime under the UCMJ, but it did not assess all UCMJ criteria for each of the additional incidents in the past year.

Unwanted Events and Types of Events Categorized as Past-Year Sexual Assault

The sexual assault section of the survey used follow-up questions to determine whether the indicated unwanted events (the six sexual assault screening items) met all UCMJ criteria for a sexual assault. Key findings on the way in which respondents answered these questions and were classified as having experienced a sexual assault can be found in Volume 2. Detailed analyses on the flow of respondents through these questions and the resulting classifications of sexual assault can be found in Volume 4.

Combining the data from the six screeners, we estimated the number of individuals who indicated having experienced any of the unwanted events described in the six screening questions (see Table 3.3) (e.g., "<u>unwanted</u> experiences in which someone <u>intentionally touched</u> private areas of your body (either directly or through clothing)").[6] These results indicate about one-half to one percentage point higher rates in each category, as compared with those who are ultimately classified as having experienced a sexual assault based on having met all of the qualifying UCMJ criteria measured in probes. For men and women, Coast Guard members were less likely to indicate these events than members of the Army, Navy, or Marine Corps. Results on any of these

Table 3.3
Estimated Percentage of Active-Component DoD and Coast Guard Service Members Who Experienced Any Type of Unwanted Event, by Gender and Service Branch

Service	Total	Men	Women
Coast Guard	1.25% (0.91–1.67)	0.85% (0.49–1.36)	3.53% (2.76–4.44)
Army	2.28%[a] (1.99–2.60)	1.73%[a] (1.41–2.10)	5.70%[a] (5.29–6.14)
Navy	3.59%[a] (3.03–4.22)	2.73[a] (2.08–3.51)	7.63%[a] (6.90–8.41)
Air Force	1.16% (1.03–1.31)	0.61% (0.47–0.79)	3.54% (3.29–3.81)
Marine Corps	2.65%[a] (2.03–3.39)	2.14%[a] (1.49–2.96)	9.07%[a] (7.80–10.47)

NOTE: 95-percent confidence intervals for each estimate are indicated in parentheses.
[a] Percentage is significantly different from Coast Guard within a column; $p < 0.05$, Bonferroni corrected.

---

[6]   Unwanted experiences include events that may not be classified as sexual assaults. To be classified as experiencing a sexual assault, the respondent must indicate they had an unwanted experience (one of the six screening questions), and they must indicate on relevant follow-up questions that the contact was abusive or sexual, and that the contact occurred by one of the types of coercion listed in Article 120 of the UCMJ.

unwanted events by gender and pay grade are presented in the Annex to Volume 3, Table A.2.

Details about how the larger sample of DoD service members answered items within this section and were ultimately classified as having experienced a sexual assault may be useful for the Coast Guard to consider, particularly regarding gender differences that are too small to be reliably estimated in the Coast Guard sample. Details can be found in Volumes 2 and 4. A summary of the main findings includes the following:

- Among men and women, unwanted, intentional touching of private areas was the most frequently indicated item. Penile and non-penile penetrative assaults were somewhat less commonly indicated. Being forced to penetrate someone else or experiencing an attempted but uncompleted penetration were rarely indicated.
- Following indication of an unwanted event, the next step in classification involved two questions designed to capture the intentional nature of the event, to conform with UCMJ definitions of sexual assaults, which require the intent be to "abuse, humiliate, harass, or degrade any person" or "arouse or gratify the sexual desire of any person" (except for penile penetration, for which verification of the offender's intentions is not required by the UCMJ). Across all screeners, men who were classified as having experienced a sexual assault in the past year were much more likely than women to indicate that the intent of the assault was to abuse or humiliate them. This gender difference in rates of describing the assault as humiliating or abusive (rather than for sexual gratification) was consistent for penetrative and non-penetrative assaults.
- Respondents who indicated that the unwanted event was abusive, humiliating, or sexual in intent were presented with a series of eight to 11 possible types of offender behaviors that were consistent with coercion or not having obtained consent and were asked to indicate whether each did or did not happen during the unwanted event. Two-thirds to almost all of respondents (across screening items) indicated that the unwanted event included either force, threats, or other forms of coercion or lack of consent. Respondents who indicated coercion or lack of consent on these items were classified as having experienced a sexual assault.
  - Among those classified as experiencing a penetrative sexual assault, the most commonly indicated forms of coercion were the offender continuing despite being told or shown that the victim was unwilling (76 percent of men and 79 percent of women) and physical force (67 percent of men and 55 percent of women). Men reported injury in 43 percent of these events and threats of injury in about one-half of the events, whereas a smaller proportion of women indicated injuries or threats.
  - Among those classified as having experienced a non-penetrative sexual assault, the most commonly indicated forms of coercion were that they either showed the offender that they were unwilling or did not consent, with about one-

quarter of cases involving the use of physical force. Injuries were less frequent in this type of assault, as was drug or alcohol incapacitation.

In summary, the data indicate that the penetrative assaults described on this survey involved more physical force and threats than the non-penetrative assaults, particularly among men, and also involved more instances of drug and alcohol incapacitation (for men and women) than non-penetrative assaults.

Reports of Sexual Assaults Prior to the Past Year

In addition to the main section of the survey, which assessed sexual assaults in the past year, all respondents were asked about experiences that happened more than a year ago, "of an abusive, humiliating, or sexual nature, and that occurred even though you did not want it and did not consent." This question also contains a definition of "did not consent." The series of questions included five items that collapsed into the same three categories used for past-year sexual assault: (1) penetrative sexual assault, (2) non-penetrative sexual assault, and (3) attempted penetration.

Lifetime Rates of Sexual Assault

By combining sexual assaults that occurred in the past year and those that occurred more than a year ago, we estimated that 4.5 percent of Coast Guard service members have experienced a sexual assault in their lifetime (Table 3.4). Compared with Coast Guard men, women are 11 times as likely to have a sexual assault during their lifetime (Table 3.4). The lifetime prevalence rates for the Coast Guard overall and for Coast Guard men are significantly lower than those in the Navy. The rates for Coast Guard women are higher than in the Air Force. Results on lifetime sexual assault by gender and pay grade can be found in the Annex to Volume 3, Table A.3.

Sexual Assault Rates Prior to Joining the Military

For those respondents indicating a lifetime sexual assault, we asked whether any sexual assault happened before they joined the military. Between 1 and 2 percent of Coast Guard service members indicated they had been sexually assaulted prior to beginning their military career (8 percent of women and less than 1 percent of men). Coast Guard service members were less likely to indicate that they experienced a sexual assault prior to joining the military than Army respondents (Table 3.5). Results on sexual assault prior to joining the military by gender and pay grade can be found in the Annex to Volume 3, Table A.4.

Table 3.4
Estimated Percentage of Active-Component DoD and Coast Guard
Service Members Who Experienced a Sexual Assault in Their
Lifetime, by Gender and Service Branch

Service	Total	Men	Women
Coast Guard	4.50% (4.01–5.03)	1.85% (1.39–2.41)	19.57% (17.94–21.28)
Army	4.45% (4.16–4.75)	2.36% (2.05–2.71)	17.46% (16.84–18.10)
Navy	6.78%[a] (6.21–7.39)	3.96%[a] (3.32–4.69)	20.03% (19.07–21.02)
Air Force	4.14% (3.95–4.34)	1.54% (1.35–1.75)	15.34%[a] (14.84–15.84)
Marine Corps	3.99% (3.38–4.69)	2.52% (1.89–3.29)	22.48% (20.73–24.31)

NOTE: 95-percent confidence intervals for each estimate are indicated in
parentheses.

[a] Percentage is significantly different from Coast Guard within a column;
$p < 0.05$, Bonferroni corrected.

Table 3.5
Estimated Percentage of Active-Component DoD and Coast Guard
Service Members Who Experienced a Sexual Assault Prior to Joining
the Military, by Gender and Service

Service	Total	Men	Women
Coast Guard	1.66% (1.36–2.01)	0.58% (0.31–0.98)	7.83% (6.74–9.02)
Army	1.83% (1.65–2.03)	0.90% (0.71–1.13)	7.69% (7.26–8.14)
Navy	2.52%[a] (2.23–2.82)	1.14% (0.84–1.50)	9.00% (8.30–9.74)
Air Force	2.03% (1.90–2.17)	0.73% (0.61–0.87)	7.62% (7.26–7.99)
Marine Corps	1.51% (1.17–1.91)	0.86% (0.53–1.31)	9.64% (8.35–11.05)

NOTE: 95-percent confidence intervals for each estimate are indicated in
parentheses.

[a] Percentage is significantly different from Coast Guard within a column;
$p < 0.05$, Bonferroni corrected.

Sexual Assault Rates Since Joining the Military

We estimated the prevalence of sexual assault during a respondent's time in the Coast Guard by combining those who were classified as having experienced a sexual assault in the past year with those who were sexually assaulted more than a year ago but after joining the Coast Guard. This is not the same as an estimate of the rates of sexual assault over the course of a career in the military, because most people in our sample have not yet completed their careers. Instead, it is a snapshot in time that provides an estimate of how many Coast Guard members currently serving have been sexually assaulted at least once since joining the Coast Guard.

Women in the Coast Guard were ten times more likely than men to report a sexual assault during their time in service (Table 3.6). Across the military services, members of the Coast Guard were less likely to indicate a sexual assault since joining the military than were those in the Navy, and more likely than those in the Air Force. Coast Guard women were more likely to indicate a sexual assault since joining the military than were Air Force women and less likely than women in the Marine Corps, whereas Coast Guard men were significantly less likely to indicate a sexual assault since joining the military than Navy men. Results on this variable by gender and pay grade can be found in the Annex to Volume 3, Table A.5.

Table 3.6
Estimated Percentage of Active-Component DoD and Coast Guard Service Members Who Experienced a Sexual Assault Since Joining the Military, by Gender and Service Branch

Service	Total	Men	Women
Coast Guard	3.83% (3.37–4.34)	1.61% (1.18–2.16)	16.45% (14.94–18.05)
Army	3.68% (3.41–3.97)	1.95% (1.66–2.28)	14.49% (13.92–15.08)
Navy	5.71%[a] (5.18–6.29)	3.37%[a] (2.77–4.07)	16.71% (15.82–17.64)
Air Force	3.10%[a] (2.93–3.27)	1.05% (0.88–1.24)	11.94%[a] (11.50–12.40)
Marine Corps	3.41% (2.83–4.07)	2.13% (1.54–2.86)	19.48%[a] (17.83–21.21)

NOTE: 95-percent confidence intervals for each estimate are indicated in parentheses.

[a] Percentage is significantly different from Coast Guard within a column; $p < 0.05$, Bonferroni corrected.

Characteristics of the Sexual Assault or the "Most Serious" of Multiple Assaults in the Past Year

Respondents who were classified as having experienced a sexual assault in the past year were asked a variety of follow-up questions describing the event. Those who reported a single event were queried about it, whereas those who reported multiple incidents in the past year were asked to reflect on the event that had the "biggest effect on you . . . that you consider to be the worst or most serious."

In the following sections, we summarize the key findings on the single or "most serious" sexual assault experienced in the prior year. Tables summarizing the items by gender, by service, and by pay grade can be found in the Annex to Volume 3, Tables A.6.a–A.38.c. To protect respondents' privacy, we do not present data on subsets with fewer than 15 respondents. The rate of sexual assault among Coast Guard men is so low that this means we do not summarize the characteristics of the sexual assaults experienced by men in the past year. The estimates are too imprecise to allow reliable inferences about the characteristics of sexual assaults against Coast Guard men.

As discussed in Chapter Two, we also omit estimates that have confidence intervals that are more than 15 percentage points above or below the estimate. We do this because with such a large confidence interval, the estimate itself is a poor indicator of the true value in the population. In such cases, we highlight the confidence interval only, which provides better guidance on where the population estimate can be predicted to lie. The effect of these rules is that we can often provide estimates for women, but not for men and not for men and women combined, due to the lack of statistical precision. Thus, we will focus the following discussion on the types and consequences of sexual assault among Coast Guard women. Confidence intervals related to men, and to the total across men and women, can be found in the Annex to Volume 3.

Type of Assault

Among Coast Guard women who had experienced a sexual assault in the past year, 51 percent indicated that they had experienced a single sexual assault. The remaining 49 percent experienced more than one sexual assault and, for the following questions, were asked to reflect on the event they considered to be most serious. Forty-one percent described penetrative assaults, 52 percent non-penetrative assaults, and the remainder described an attempted penetrative assault. See the Annex to Volume 3, Tables A.6.a–A.6.c and Tables A.8.a–A.8.c.

Description of Offender(s)

In the majority of assaults against Coast Guard women, the offender(s) were a man or men (93 percent). Most respondents indicated that there was a single offender, and most of the offender(s) were known to the respondent (92 percent). However, few were intimate partners or family members. A substantial number of respondents said the offender was a "friend or acquaintance" (52 percent). The majority of respondents indicated that the assailant was in the military or, if they were assaulted by a group, that at least one assailant was in the military (77 percent). This rate is significantly lower than the proportion of women assaulted by a member of the military in each of the DoD services other than the Air Force, in which 82 percent of women who experienced sexual assault in the past year said the offender was a service member. When the offender was a member of the military, often it was someone of higher rank (95% CI: 38–72). The offender was a civilian or contractor working for the military in 3 percent of the sexual assaults described. See the Annex to Volume 3, Tables A.7.a–A.7.c and Tables A.9.a–A.15.c for additional details.

Description of Assault Location and Circumstances

Consistent with the identities of offenders described above, a substantial number of Coast Guard women indicated that the event occurred on a military installation or ship (95% CI: 22.2–50.1); during the work day or duty hours (18 percent); and/or while on temporary duty (TDY)/temporary additional duty (TAD), at sea, or during field exercises/alerts (27 percent). Five percent indicated it occurred while deployed to a combat zone. Other types of military training activities were more rarely indicated, perhaps because low numbers of respondents participated in them. See the Annex to Volume 3, Tables A.16.a–A.16.c, for details.

In terms of contextual factors, about one-half of Coast Guard women (51 percent) indicated the assault occurred when "out with friends or at a party," whereas 17 percent indicated it happened while at work. As such, past-year sexual assaults are significantly less likely to occur at work for women in the Coast Guard than for those in the Army and Navy. Fifteen percent of Coast Guard women who were sexually assaulted indicated that they were in their own home or quarters; 19 percent indicated they were in someone else's home or quarters. Among Coast Guard women, 3 percent indicated they would describe the event as hazing. See the Annex to Volume 3, Tables A.17.a–A.18.c, for details.

Thirty percent of women who were sexually assaulted indicated that the offender(s) sexually harassed them before the assault, and 15 percent indicated the offender harassed them after the assault took place. (See the Annex to Volume 3, Tables A.19.a–A.19.c, for details.) We also examined classification of sexual harassment on the survey. Among women who were classified as having experienced sexual harassment in the past year (see Chapter Four of this volume), 13.16 percent (95% CI: 8.80–18.66) also experienced a sexual assault during that year. In contrast, rates of sexual assault were much lower among those who did not experience sexual harassment (95% CI: 0.48–1.69). This strong association is attributable, in part, to the fact that sexual assaults by coworkers could also be counted as sexual harassment.

Sixty-three percent of Coast Guard women who were sexually assaulted indicated that they had been drinking at the time of the assault, and 8 percent indicated that they may have been given a drug without their knowledge or consent. A majority indicated that the offender(s) had been drinking alcohol at the time of the assault (62 percent). In all, 75.91 percent (95% CI: 63.20–85.97) indicated that either the respondent, the offender, or both had been drinking. See the Annex to Volume 3, Tables A.20.a–A.20.c, for details.

Assaults that involved another service member, someone who works for the military, or that occurred in a military location or at a military function accounted for 79.93 percent of all assaults against Coast Guard women (95% CI: 67.45–89.24).

Consequences of the Past-Year Assault

Respondents also answered questions about specific impacts of the single or most serious sexual assault that occurred in the past year. Forty-six percent of Coast Guard women indicated the assault made it hard to do their work, 23 percent indicated that they took a sick day or other leave because of the event, and 6 percent indicated they requested a transfer or change of duty assignment. At least one-fifth of female respondents (95% CI: 20–48 percent) indicated that the assault made them want to leave the military. About one-half (51 percent) indicated that the assault damaged their personal relationships. See the Annex to Volume 3, Tables A.21.a–A.21.c, for details.

Telling Others/Reporting Past-Year Assault

Two-thirds of female respondents who were sexually assaulted indicated that they told someone about the assault. About 3 out of 5 (59 percent) talked about it with a friend or family member. The most common military resources contacted by women who had been assaulted were counselors/therapists (21 percent), sexual assault prevention and response victim advocates (19 percent), sexual assault response coordinators (16 percent), and chaplains or religious leaders (12 percent). Rarely did respondents indicate that they contacted resources outside the military system, such as civilian law enforcement. We asked respondents who talked to each resource about how satisfied they were with the experience and generally found satisfaction to be high, with large majorities indicating they were satisfied or very satisfied with the experience.

Eighteen percent of women who experienced sexual assault filed an official report about it to the military.[7] We also asked all respondents who experienced a sexual assault if they signed a DD Form 2910[8] for an assault in the past year. These Victim Prefer-

---

[7] Two types of official reports are possible. *Restricted* reports allow people to get information, collect evidence, and receive medical treatment and counseling without disclosing the details of the assault to an investigative authority (and therefore, without initiating an investigation). *Unrestricted* reports start an official investigation, in addition to allowing the support services available in restricted reporting.

[8] DD Form 2910, also known as the Victim Preference Reporting Statement, is a document on which a sexual assault victim chooses whether to make a restricted or unrestricted report of the assault to the military. However, if an individual other than the victim reports the assault, an independent investigation may be initiated by a military criminal investigative organization.

ence Reporting Statements serve as the basis for official DoD statistics on sexual assault reporting. The survey included a link to an image of the form to enhance recall. Fifteen percent of women who were sexually assaulted in the past year indicated they had signed or initialed this form, and an additional 7 percent indicated they were not sure.

Thirteen percent of women who experienced sexual assault were interviewed by military police or a criminal investigator about the case, and 2 percent indicated the suspect had been arrested or charged with a crime by the date the survey was fielded. We asked several questions about the status of the criminal case, but the sample size was too small to analyze these responses. Given that these assaults took place between 0 and 12 months ago, criminal investigations and prosecutions may have been in the early stages of the UCMJ process for many assaults.

Among those who did not make an official report, we asked for their reasons for not reporting the incident. To identify important points of intervention, we asked participants to indicate their primary reason for not reporting. The most frequently indicated primary reasons for not reporting were that the respondent "wanted to forget about it and move on," "took other actions to handle the situation," "thought it was not serious enough to report," or "felt partially to blame." Seventy-six percent of past-year female assault victims indicated they would make the same choice about reporting if they had to make the decision again. There were too few cases to determine if this percentage varies as a function of whether the respondent officially reported the assault or not. Members of the Coast Guard were significantly less likely than those in the DoD services to say they did not make a report because they feared they would be viewed as gay, lesbian, bisexual, or transgender.

See the Annex to Volume 3, Tables A.22.a–A.34.c and Tables A.37.a–A.38.c, for additional details on this topic.

Perceived Retaliation or Negative Career Actions

The survey included four items asking those who experienced a sexual assault if they perceived they experienced retaliation or negative career actions related to the sexual assault. Responses to the individual items and sources of retaliation can be found in the Annex to Volume 3, Tables A.35.a–A.36.c, and ranged from a low of 2 percent for being punished to a high of 29 percent for social retaliation. Thirty-two percent of women who were sexually assaulted (31.51 percent; 95% CI: 19.33–45.89) reported at least one of these four types of retaliation or negative career actions. Due to a small sample size, we were unable to examine the rate of retaliation or adverse actions among those who made an official report.

Summary

In the year prior to the survey fielding, 3 percent of active-component Coast Guard women and 0.3 percent of men experienced at least one sexual assault, as defined in the UCMJ. About one-half of individuals who experienced a sexual assault in the past year experienced more than one such event. Due to low numbers of men in the sample who had experienced a sexual assault in the past year, description of the types of events experienced focuses on those events experienced by Coast Guard women. The types and patterns of assaults showed substantial variability, but 80 percent of the assaults against Coast Guard women occurred in a military context (e.g., at a military installation, during work hours, by an offender in the military). Findings suggest that these assaults affected many Coast Guard women in terms of personal relationships, work productivity, and a desire to leave the military. About two-thirds of Coast Guard women who were assaulted told someone about it, and 18 percent made an official report. Among those Coast Guard women who talked to someone about the assault, they were generally satisfied with the experience.

Sexual Harassment and Gender Discrimination Findings: Coast Guard Active Component

Coreen Farris, Terry L. Schell, Lisa H. Jaycox,
Amy E. Street, Dean G. Kilpatrick, and Terri Tanielian

In this chapter, we provide estimates of the proportion of the active-component Coast Guard members who experienced one of two forms of sexual harassment (a sexually hostile work environment or *quid pro quo* harassment) or gender discrimination in the past year. According to military directives, both sexual harassment and gender discrimination are sex-based MEO violations. For those who experienced sexual harassment or gender discrimination in the past year, we also report the characteristics of the events and the offender,[1] the effect on workplace productivity and intentions to stay on active duty, disclosure choices, responses to official reports, and barriers to reporting among those who chose not to do so. Findings will be of interest to Coast Guard leaders, policymakers, and the public.

Prevalence of Sexual Harassment and Gender Discrimination

Our measures of sexual harassment and gender discrimination assessed a number of specific types of MEO violations. All of the violations focused on the military workplace by querying about inappropriate workplace behaviors committed by "someone from work." We used the phrase "someone from work" rather than "coworker" to ensure that respondents included all work contacts, not just those they perceived as peers. We asked respondents to consider any person they have contact with as part of their military duties, and reminded them that this person could be a supervisor, above or below them in rank, a civilian employee or contractor, and could be in their unit or other units.

---

[1] We use the term *offender(s)* to refer to the person or people who sexually harassed or discriminated against the respondent. We acknowledge that not all forms of sexual harassment and gender discrimination are necessarily illegal, but prefer offender, as more readily interpretable to all readers, over the term *source* often used in the academic literature.

The *sexually hostile work environment* measure was designed to capture a type of sexual harassment that includes sexual language, gestures, images, or behaviors that offend or anger service members. These inappropriate workplace events are categorized as a hostile workplace violation if the offensive behavior was either persistent (i.e., the respondent indicated the behavior continued even after the person[s] knew that it was upsetting to others) or is described by the respondent as severe (i.e., the behavior was so severe that most service members would find it patently offensive). Table 4.1 shows that this type of sexual harassment is faced by some active-component Coast Guard members (6 percent) and is more common for women than for men. We estimate that one-fifth of women experienced upsetting or offensive sexual behavior in the past year that DoD directives would define as an unlawful form of discrimination that deprives service members of their rights to equal opportunities in the military.[2] Men and women in the Coast Guard were less likely than members of the Army, Navy, or Marine Corps to experience a sexually hostile work environment in the past year. However, women

Table 4.1
Estimated Percentage of Active-Component Coast Guard and DoD Service Members Who Experienced a Sexually Hostile Work Environment in the Past Year, by Gender and Service Branch

Service	Total	Men	Women
Coast Guard	6.00% (5.22–6.85)	3.74% (2.94–4.68)	19.15% (17.05–21.39)
Army	9.75%[a] (9.01–10.53)	7.65%[a] (6.81–8.56)	22.87%[a] (21.92–23.84)
Navy	11.73%[a] (10.60–12.94)	8.34%[a] (7.02–9.81)	27.71%[a] (26.21–29.26)
Air Force	4.96% (4.56–5.38)	3.26% (2.80–3.77)	12.32%[a] (11.72–12.95)
Marine Corps	7.68% (6.41–9.13)	6.11%[a] (4.76–7.70)	27.19%[a] (24.68–29.80)

NOTE: 95-percent confidence intervals for each estimate are included in parentheses.

[a] Percentage is significantly different from Coast Guard within a column; $p < 0.05$, Bonferroni corrected.

---

[2] The RAND instrument to measure sex-based MEO violations was designed to parallel the definition of these violations as specified in Department of Defense Directive 1350.2 (see Morral, Gore, and Schell, 2014). We employed the same approach to measuring sexual harassment and gender discrimination, without revisions, to the Coast Guard study described here. The Coast Guard defines sexual harassment and gender discrimination in the *Coast Guard Civil Rights Manual* (U.S. Coast Guard, 2010, p. 2-C.9), which differs slightly from DoDD 1350.2. However, we do not believe these modest differences affect the interpretation of the study results or their applicability to the Coast Guard.

in the Coast Guard were more likely than Air Force women to experience a sexually hostile work environment.

The second sexual harassment measure, sexual *quid pro quo*, a Latin phrase meaning "this for that," identifies incidents in which someone used their power or influence within the Coast Guard to attempt to coerce sexual behavior. These inappropriate workplace events are categorized as a sexual harassment violation if respondents indicated they had direct evidence that a workplace benefit or punishment was contingent on a sexual behavior. Hearsay or rumor was not considered sufficient evidence to categorize an event as a *quid pro quo* violation. Unlike sexually hostile work environment violations, this form of sexual harassment was comparatively rare (Table 4.2). We estimate that less than 1 percent of active-component Coast Guard members experienced a *quid pro quo* violation in the past year and that between 10 and 50 active-component Coast Guard women had such experiences in the past year. Active-component men in the Coast Guard were less likely than men in the Army and Air Force (but not significantly less likely than men in the Navy or Marine Corps) to experience a sexual *quid pro quo* violation in the past year. Coast Guard women were less likely than women in the Army, Navy, and Marine Corps to experience a *quid pro quo* violation (but at similar risk to women in the Air Force).

In the Coast Guard, *quid pro quo* events are much rarer than those reflecting a sexually hostile work environment, but they still represent a particularly serious category of offense. Because Coast Guard leaders have great authority over members' lives—more so than supervisors in the civilian workplace—this type of misuse of

Table 4.2
Estimated Percentage of Active-Component Coast Guard and DoD Service Members Who Experienced Sexual *Quid Pro Quo* in the Past Year, by Gender

Service	Total	Men	Women
Coast Guard	0.07% (0.02–0.19)	0.00% (0.00–0.20)	0.50% (0.23–0.93)
Army	0.65%[a] (0.49–0.84)	0.41%[a] (0.25–0.64)	2.12%[a] (1.79–2.49)
Navy	0.80%[a] (0.43–1.38)	0.50% (0.12–1.34)	2.22%[a] (1.70–2.85)
Air Force	0.14% (0.10–0.20)	0.06%[a] (0.03–0.12)	0.50% (0.37–0.65)
Marine Corps	0.50% (0.16–1.20)	0.37% (0.05–1.26)	2.12%[a] (1.31–3.25)

NOTE: 95-percent confidence intervals for each estimate are included in parentheses.

[a] Percentage is significantly different from Coast Guard within a column; $p < 0.05$, Bonferroni corrected.

their authority is a significant concern. In some cases, these acts are also likely to be crimes (e.g., UCMJ Articles 92, 93, 133, and 134), not purely MEO violations. Thus, although rare, it will be valuable to monitor these offenses over time to assess whether the prevalence of these offenses is being reduced.

The two measures we have discussed thus far, sexually hostile work environment and sexual *quid pro quo*, together constitute the legal construct of sexual harassment. Thus, our sexual harassment measure (Table 4.3) includes all targets of either of these subtypes of sexual harassment. Approximately 6 percent of active-component Coast Guard members were classified as experiencing some form of sexual harassment in the past year, which corresponds to 2,350 members (95% CI: 2,050–2,690). The overall measure of sexual harassment may not be as descriptively useful as its components, however, because it is dominated by the more common form of harassment (sexually hostile work environment). A comparison between Table 4.3 and Table 4.1 shows that the aggregate rate of sexual harassment is almost identical to the rate of sexually hostile work environment, which means that the vast majority of individuals who indicated sexual *quid pro quo* in the past year also indicated being sexually harassed in a sexually hostile work environment. These results also suggests that sexually hostile work environments may put members at higher risk for sexual *quid pro quo* overtures; that is, the vast majority of those describing *quid pro quo* experiences also described having

Table 4.3
Estimated Percentage of Active-Component Coast Guard and DoD Service Members Who Experienced Sexual Harassment in the Past Year, by Gender and Service Branch

Service	Total	Men	Women
Coast Guard	6.02% (5.24–6.88)	3.75% (2.94–4.69)	19.19% (17.09–21.43)
Army	9.80%[a] (9.05–10.58)	7.67%[a] (6.83–8.58)	23.07%[a] (22.12–24.05)
Navy	11.78%[a] (10.65-12.99)	8.37%[a] (7.05–9.84)	27.82%[a] (26.31–29.36)
Air Force	4.99% (4.60–5.42)	3.29% (2.82–3.80)	12.43%[a] (11.82–13.07)
Marine Corps	7.69% (6.42–9.14)	6.11%[a] (4.76–7.70)	27.30%[a] (24.79–29.92)

NOTE: 95-percent confidence intervals for each estimate are included in parentheses.

[a] Percentage is significantly different from Coast Guard within a column; $p < 0.05$, Bonferroni corrected.

experienced a sexually hostile workplace in the past year.[3] Both men and women in the Coast Guard were at lower risk for sexual harassment than were men and women in the Army, Navy, or Marine Corps. However, women serving in the Coast Guard were at higher risk for sexual harassment than were women serving in the Air Force.

The *gender discrimination* measure assesses incidents in which the respondent indicated that he or she heard derogatory gender-related comments or was mistreated on the basis of his or her gender. For inappropriate workplace events to be categorized as a gender discrimination violation, respondents had to indicate that the mistreatment harmed their military career (e.g., adversely affected their evaluation, promotion, or assignment). We estimate that gender discrimination affected approximately 1 in 8 active-component Coast Guard women in the past year and 1 in 95 men (Table 4.4). These rates correspond to 1,020 Coast Guard members (95% CI: 830–1,250) who experienced gender discrimination in the past year. Men in the Coast Guard were at lower risk for gender discrimination than men in the Army and Navy. Coast Guard women were at lower risk than Army, Navy, and Marine Corps women, but were at higher risk relative to women serving in the Air Force.

Table 4.4
Estimated Percentage of Active-Component Coast Guard and DoD Service Members Who Experienced Gender Discrimination in the Past Year, by Gender and Service Branch

Service	Total	Men	Women
Coast Guard	2.62% (2.12–3.19)	1.05% (0.59–1.72)	11.75% (10.12–13.55)
Army	3.86%[a] (3.54–4.21)	2.11%[a] (1.77–2.49)	14.80%[a] (14.02–15.61)
Navy	4.65%[a] (4.07–5.28)	2.52%[a] (1.89–3.27)	14.65%[a] (13.50–15.86)
Air Force	1.95% (1.78–2.13)	0.86% (0.70–1.04)	6.69%[a] (6.23–7.17)
Marine Corps	1.97% (1.62–2.38)	0.87% (0.60–1.23)	15.59%[a] (13.65–17.70)

NOTE: 95-percent confidence intervals for each estimate are included in parentheses.

[a] Percentage is significantly different from Coast Guard within a column; $p < 0.05$, Bonferroni corrected.

---

[3] In the field of epidemiology, the association between a risk factor and an outcome is often described in terms of a risk ratio, or the ratio of the probability of an event occurring in an exposed group relative to that in a group not exposed. Risk ratios of five or ten are almost always considered large (McMahon and Pugh, 1970). Our results suggest the risk ratio of *quid pro quo* as a function of hostile work environment is 121 in the Coast Guard sample and 101 in the much larger DoD sample.

The concept of gender discrimination is particularly challenging to assess in a self-report survey. Unlike sexual harassment, many forms of gender discrimination occur without the victim's awareness. For example, a service member who does not receive a valuable assignment because of his or her gender may never be aware that his or her career had been influenced by factors other than merit. Because these estimates are based on self-reports, they cannot count incidents in which discrimination occurred without the respondent knowing, and we cannot estimate how common these hidden cases of discrimination may be. On the other hand, respondents may sometimes attribute mistreatment to their gender when there were other legitimate causes of their adverse work experience. In spite of these interpretational difficulties, the fact that 1 out of every 8 women perceived themselves to be treated unfairly by the Coast Guard because of their gender represents an important problem.

Given that both sexual harassment and gender discrimination are MEO violations, leaders will want to know the proportion of Coast Guard members who experienced either of these events in the past year. Table 4.5 and Table B.5 in the Annex to Volume 3 provide this information. Note that the totals for members who experienced either sexual harassment or gender discrimination are noticeably higher than the total for either experience individually. This suggests that a substantial proportion of those who experienced gender discrimination did not also experience sexual harassment (see also Figure 4.3). Because this measure combines several distinct phenomena that are likely to be affected by different types of policy or educational interventions, this combined measure may not be ideal for evaluating Coast Guard progress on achieving key

Table 4.5
Estimated Percentage of Active-Component Coast Guard and DoD Service Members Who Experienced Sexual Harassment or Gender Discrimination in the Past Year, by Gender and Service Branch

Service	Total	Men	Women
Coast Guard	7.28% (6.40–8.23)	4.51% (3.60–5.57)	23.32% (21.10–25.66)
Army	11.30%[a] (10.54–12.10)	8.53%[a] (7.67–9.45)	28.62%[a] (27.61–29.64)
Navy	13.56%[a] (12.39–14.79)	9.61%[a] (8.25–11.11)	32.16%[a] (30.62–33.72)
Air Force	6.05% (5.64–6.48)	3.84% (3.36–4.37)	15.66%[a] (14.99–16.35)
Marine Corps	8.51% (7.21–9.95)	6.65% (5.28–8.25)	31.43%[a] (28.85–34.11)

NOTE: 95-percent confidence intervals for each estimate are included in parentheses.

[a] Percentage is significantly different from Coast Guard within a column; $p < 0.05$, Bonferroni corrected.

MEO goals. Even relatively substantial changes in gender discrimination or sexual *quid pro quo* over time may be difficult to detect in this aggregate measure.

Relationship Between Pay Grade and Sexual Harassment

In general, the differences in the rates of sexual harassment across pay grade were not large. No significant differences in the rate of sexual harassment among men across pay grade emerged. Among women, a lower proportion of senior enlisted (16 percent) than junior enlisted (23 percent) Coast Guard members experienced sexual harassment in the past year. See Figure 4.1 and the Annex to Volume 3, Tables B.1–B.3, for complete results.

Relationship Between Pay Grade and Gender Discrimination

Rates of gender discrimination were similar across pay grades. Approximately the same percentage of senior enlisted women (10 percent) and junior enlisted women (11 percent) were categorized as experiencing gender discrimination in the past year. A similar proportion of senior female officers (17 percent) and junior female officers (15 percent) experienced gender discrimination in the past year. The apparent differences (visually) between officers and enlisted women were not statistically significant. The same

Figure 4.1
Percentage of Active-Component Coast Guard Members Who Experienced Sexual Harassment in the Past Year, by Gender and Pay Grade

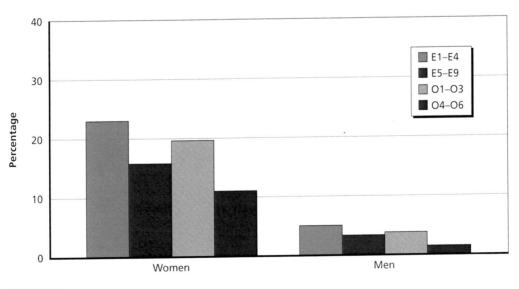

was true among men, where pay grade had no significant effect on the likelihood of experiencing gender discrimination in the past year. See Figure 4.2 and the Annex to Volume 3, Table B.4, for complete results.

Co-Occurrence of Sexual Harassment and Gender Discrimination

Figure 4.3 provides an illustration of the extent to which Coast Guard members who experienced sexual harassment (sexually hostile work environment or sexual *quid pro quo)* also experienced gender discrimination in the past year. Of Coast Guard members who were sexually harassed, nearly one-quarter also experienced gender discrimination (23 percent). Of those who experienced gender discrimination, one-half were also sexually harassed (52 percent).

Inappropriate Workplace Behaviors

The RAND assessment of sexual harassment and gender discrimination began with a series of questions to assess inappropriate workplace behaviors. For those who have experienced an inappropriate workplace behavior, the survey relied on follow-up questions to assess whether the behavior would meet criteria for an MEO violation. Although, for some respondents, the inappropriate workplace behaviors were not

Figure 4.2
Percentage of Active-Component Coast Guard Members Who Experienced Gender Discrimination in the Past Year, by Gender and Pay Grade

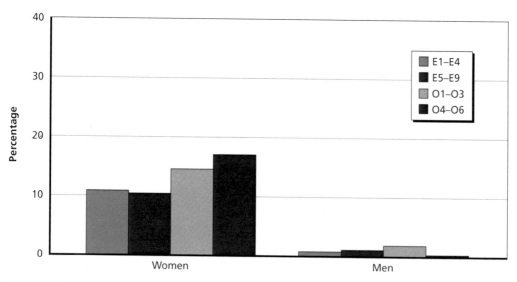

Figure 4.3
**Proportion of Service Members Experiencing Sexual
Harassment and Gender Discrimination and the Relative
Overlap Between These Military Equal Opportunity
Violations**

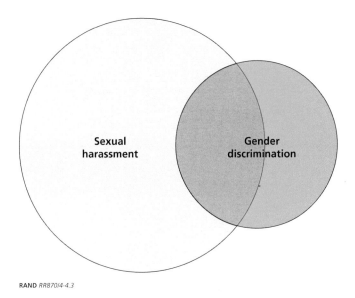

RAND *RR870/4-4.3*

ultimately characterized as sexual harassment or gender discrimination, many Coast Guard leaders will nonetheless be interested in these behaviors as possible precursors to more serious violations and as evidence of poor discipline in the workplace. In this section, we describe the past-year prevalence of each surveyed inappropriate workplace behavior. Further details about the instrument design and performance are available in Volumes 1 and 4.

Table 4.6 presents the proportion of individuals who indicated that they experienced any of the 15 inappropriate workplace behaviors in the past year (whether or not they also met persistence, severity, direct evidence, or harm to career criteria assessed via follow-up questions). With the exception of two more-rare behaviors ("take or share sexually suggestive pictures or videos of you" and "feel like you would get punished or treated unfairly in the workplace if you did not do something sexual"), women were more likely than men to have experienced each. In the most extreme differentiation between the genders, women were 22 times more likely than men to indicate that someone from work had made repeated attempts to establish an unwanted romantic or sexual relationship that the respondent found offensive.

As seen in Table 4.6, some inappropriate workplace behaviors were quite common. For example, 1 out of every 4 Coast Guard women (25 percent) indicated that someone from work had "mistreated, ignored, excluded, or insulted you because you are a woman." Others were more rare; for example, 4 out of every 1,000 Coast Guard

Table 4.6
Estimated Percentage of Active-Component Coast Guard Members Who Experienced Each Type of Inappropriate Workplace Behavior in the Past Year, by Gender

	Men	Women
Repeatedly tell sexual "jokes" that made you uncomfortable, angry, or upset?	3.2% (2.49–4.06)	12.1% (10.31–14.00)
Embarrass, anger, or upset you by repeatedly suggesting that you do not act like a [man/woman] is supposed to?	2.9% (2.22–3.74)	6.0% (4.79–7.49)
Repeatedly make sexual gestures or sexual body movements that made you uncomfortable, angry, or upset?	1.4% (0.90-2.00)	4.1% (3.01–5.46)
Display, show, or send sexually explicit materials like pictures or videos that made you uncomfortable, angry, or upset?	1.0% (0.61–1.44)	3.4% (2.51–4.55)
Repeatedly tell you about their sexual activities in a way that made you uncomfortable, angry, or upset?	1.7% (1.18–2.38)	6.7% (5.36–8.34)
Repeatedly ask you questions about your sex life or sexual interests that made you uncomfortable, angry, or upset?	1.1% (0.67–1.57)	5.4% (4.21–6.92)
Make repeated sexual comments about your appearance or body that made you uncomfortable, angry, or upset?	1.0% (0.57–1.54)	7.0% (5.66–8.63)
Either take or share sexually suggestive pictures or videos of you when you did not want them to? AND Did this make you uncomfortable, angry, or upset?	0.4% (0.17–0.73)	0.7% (0.37–1.32)
Make repeated attempts to establish an unwanted romantic or sexual relationship with you? AND Did these attempts make you uncomfortable, angry, or upset?	0.2% (0.05–0.50)	4.4% (3.36–5.62)
Intentionally touch you in a sexual way when you did not want them to?	0.1% (0.04–0.35)	1.8% (1.09–2.87)
Repeatedly touch you in any other way that made you uncomfortable, angry, or upset?	1.3% (0.82–2.08)	5.0% (3.88–6.44)
Made you feel as if you would get some workplace benefit in exchange for doing something sexual?	0.1% (0.01–0.30)	1.0% (0.52–1.61)
Made you feel like you would get punished or treated unfairly in the workplace if you did not do something sexual?	0.05% (0.00–0.27)	0.4% (0.15–0.74)
Did you hear someone from work say that [men/women] are not as good as [women/men] at your particular job, or that [men/women] should be prevented from having your job?	1.2% (0.68–1.89)	18.5% (16.53–20.68)
Do you think someone from work mistreated, ignored, excluded, or insulted you because you are a [man/woman]?	2.5% (1.76–3.35)	25.4% (23.15–27.76)

NOTE: 95-percent confidence intervals for each estimate are included in parentheses.

women (0.4 percent) indicated that someone from work had offered to withhold a workplace punishment in exchange for doing something sexual.

Types of Sexual Harassment and Gender Discrimination Violations

Next, we review the proportion of active-component Coast Guard members who—for each inappropriate workplace behavior—were categorized as experiencing sexual harassment or gender discrimination as defined by legal precedent or DoD directives.

For the inappropriate hostile workplace behaviors, respondents were categorized as having experienced a sexually hostile work environment violation if they also indicated that the behavior continued even after the offenders were aware that someone wanted them to stop (persistence) or if the respondent believed the behavior was severe enough that most people of the same gender in the military would be offended if it had happened to them (severity/reasonable person standard). The percentages of men and women who experienced each type of event are summarized in Table 4.7; in the Annex to Volume 3, Table B.6.a provides further details. Note that this summary is for those who met the legal or DoD standard for sexual harassment, as opposed to the inappropriate behaviors summarized in Table 4.6, which included all events—those that did and did not rise to the level of a violation.

For the inappropriate *quid pro quo* workplace behaviors, respondents were categorized as having experienced a *quid pro quo* violation if they had direct evidence that an offer or exchange occurred. Those who had only indirect evidence (i.e., heard rumors or inferred it from the person's personality) were not included among those who experienced a *quid pro quo* violation.

Finally, for inappropriate gender discrimination behaviors, respondents were categorized as having experienced a gender discrimination violation if they also indicated that the person's behavior had directly harmed their career.

Table 4.7 summarizes and Figure 4.4 illustrates the estimated percentage of Coast Guard men and women who were the target of workplace behaviors that met our sexual harassment (sexually hostile work environment or *quid pro quo* violation) or gender discrimination criteria. In the figure, types of violations were ordered from the most to least prevalent among women. The most common violations for women were being mistreated due to their gender with a negative impact on their career (11 percent), offensive sexual jokes in the workplace that were persistent or severe (10 percent), and someone from work making discriminatory comments about women that negatively impacted their career (7 percent). The most common violations for men were offensive sexual jokes in the workplace that were persistent or severe (2 percent), being accused of not acting according to men's gender role in a persistent or severe manner (1 percent), and being mistreated due to their gender with a negative impact on their career (1 percent). With the exception of someone taking or sharing sexually

Table 4.7
Estimated Percentage of Active-Component Coast Guard Members Who Experienced Each Type of Sexual Harassment (Hostile Workplace or *Quid Pro Quo*) or Gender Discrimination Violation in the Past Year

	Men (%)	Women (%)
Sexually hostile work environment violations	**3.7**	**19.2**
Repeatedly tell sexual "jokes" <u>that made you uncomfortable, angry, or upset</u>? Events were persistent or severe.[a]	1.5	10.2
Embarrass, anger, or upset you by repeatedly suggesting that you do not act like a [man/woman] is supposed to? Events were persistent or severe.[a]	1.3	5.5
Repeatedly make sexual gestures or sexual body movements (for example, thrusting their pelvis or grabbing their crotch) that made you uncomfortable, angry, or upset? Events were persistent or severe.[a]	0.7	3.5
Display, show, or send sexually explicit materials like pictures or videos that made you uncomfortable, angry, or upset? Events were persistent or severe.[a]	0.4	2.9
Repeatedly tell you about their sexual activities in a way that made you uncomfortable, angry, or upset? Events were persistent or severe.[a]	0.8	4.9
Repeatedly ask you questions about your sex life or sexual interests that made you uncomfortable, angry, or upset? Events were persistent or severe.[a]	0.6	4.4
Make repeated sexual comments about your appearance or body that made you uncomfortable, angry, or upset? Events were persistent or severe.[a]	0.5	5.8
Either <u>take or share</u> sexually suggestive pictures or videos of you when you did not want them to? AND Did this make you uncomfortable, angry, or upset? Events were persistent or severe.[a]	0.1	0.5
Make <u>repeated</u> attempts to establish an <u>unwanted</u> romantic or sexual relationship with you? AND Did these attempts make you uncomfortable, angry, or upset? Events were persistent or severe.[a]	0.2	3.8
Intentionally touch you in a sexual way when you did not want them to? Categorized as severe without additional follow-up questions.	0.1	1.8
Repeatedly touch you in any other way that made you uncomfortable, angry, or upset? Events were persistent or severe.[a,b]	0.8	5.7
***Quid pro quo* violations**	**0.0**	**0.5**
Direct evidence of a workplace benefit in exchange for doing something sexual?[c]	0.0	0.4
Direct evidence of a threat of <u>punishment or unfair treatment in the workplace</u> if you did <u>not</u> do something sexual?[c]	0.0	0.3

Table 4.7—Continued

	Men (%)	Women (%)
Gender discrimination violations	**1.1**	**11.8**
Perceived harm to military career based on hearing someone from work say that [men/women] are <u>not</u> as good as [women/men] at your particular job, or that [men/women] should be prevented from having your job.[d]	0.4	6.8
Perceived harm to military career because someone from work mistreated, ignored, excluded, or insulted you because you are a [man/woman]?[d]	1.0	10.8

[a] Follow-up questions established that the event(s) were persistent (the behavior continued even after the person was aware that someone wanted them to stop) or severe (most people of the same gender in the military would be offended if it had happened to them).

[b] Respondents who were touched in a sexual way are also categorized in this more-inclusive "any touching" category. For this reason, the percentage of those classified as experiencing this type of sexual harassment is larger than the percentage who indicated they experienced this particular type of inappropriate workplace behavior (which was not asked of those who indicated SH10, "Intentionally touch you in a sexual way when you did not want them to?").

[c] Follow-up questions established that the respondent had direct evidence of an offer (rumors or the respondent's inference based on the person's personality were not adequate to categorize the event as a *quid pro quo* violation).

[d] A follow-up question assessed whether the event(s) harmed the respondent's military career (e.g., hurt an evaluation/fitness report, affected promotion or next assignment).

suggestive pictures/videos of the respondent, men were significantly less likely than women to experience each type of sexual harassment and gender discrimination violation. Tables B.6.b–B.6.c in the Annex to Volume 3 provide detailed analyses of service and pay grade differences.

Many Coast Guard members indicated that they experienced more than one of the fifteen measured forms of sexual harassment and gender discrimination violations. For those who had at least one experience that rose to the level of a violation, the average number of sexual harassment and gender discrimination types experienced in the past year was 2.9 for women (SE = 0.13; Min = 1; Max = 15) and 1.9 for men (SE = 0.13; Min = 1; Max = 9). This convergence of events is important to keep in mind when interpreting the values in Table 4.7 and Figure 4.4. Many of the individuals who are classified as having a certain type of sexual harassment or gender discrimination experience will also have experienced other types of events.

Self-Identification of Events as Sexual Harassment

We asked Coast Guard members who were categorized as having experienced sexual harassment whether they believed the events they experienced were sexual harassment. Women (32 percent) were less likely than men (67 percent) to deny that their experi-

Figure 4.4
Percentage of Active-Component Coast Guard Women and Men Who Experienced Each Type of Sexual Harassment and Gender Discrimination Violation in the Past Year

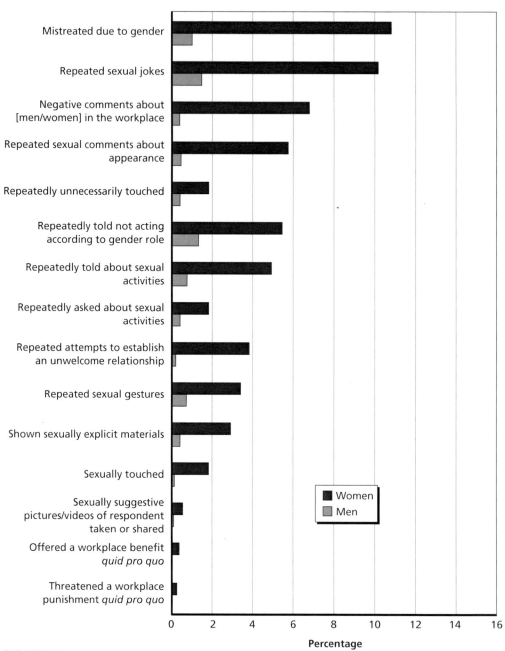

ences constituted sexual harassment (see the Annex to Volume 3, Tables B.7.a–B.7.c). The relatively large proportion of individuals who did not self-label their experiences as harassment, which we classified as actually meeting Coast Guard sexual harassment criteria, may reflect a number of issues. First, educational efforts to teach Coast Guard members the boundaries of professional workplace behaviors and the definition of sexual harassment seem not to have been fully successful. This appears to be particularly relevant to men, for whom the discrepancy is larger. Alternatively, some respondents may feel uncomfortable characterizing their own experiences as sexual harassment or may be hesitant to paint their Coast Guard workplace in a poor light on a survey. In either case, the survey data indicate that there is a discrepancy between having experienced events that we classified as meeting Coast Guard criteria for sexual harassment and being capable of self-identifying those events as sexual harassment. Moving forward, the degree of mismatch could be a potential metric to assess the success of Coast Guard sexual harassment education and stigma-reduction efforts.

Description of Past-Year Sexual Harassment or Gender Discrimination

All respondents who had experiences consistent with legal and military definitions of sexual harassment or gender discrimination were asked a series of questions that assessed the characteristics of these events, their disclosure choices, the system response to disclosed events, and barriers to reporting among those who chose not to disclose their experiences. Some respondents who had experienced sexual harassment or gender discrimination in the past year indicated that it had occurred in different situations and was committed by different people (37 percent; 95% CI: 31.09–44.16). These individuals responded to all subsequent questions while considering the situation that had the "biggest effect" on them, the one they considered "to be the worst or most serious." For this reason, the descriptions that follow are representative of a target's single or most serious sexual harassment or gender discrimination experience. It is possible that an account of all situations (rather than victims' choices of their worst experience when multiple occurred) would be different than the description reported here. For example, if victims select coworker-perpetrated events more often than supervisor-perpetrated events as their most serious situation, then we would expect the proportion of supervisors represented in all sexual harassment situations to be higher than the proportion of supervisors we measured for single or worst sexual harassment situations. The following descriptive statistics are limited to those who experienced sexual harassment or gender discrimination as defined by the Coast Guard (unless otherwise noted).

Characteristics of the Offender

Thirty-three percent of targets indicated that there was more than one person who harassed or discriminated against them (see Table 4.8 and the Annex to Volume 3,

Table 4.8
Characteristics of the Situation and Offenders

	Total	Men	Women
Number of offenders			
Individual	67%	65%	69%
Group	33%	35%	31%
Gender of the offender(s)			
Man or men only	81%	70%	93%
Woman or women only	11%	18%	2%
Mix of men and women	9%	12%	5%
Duration of situation			
One time	23%	25%	21%
About one week	8%	8%	8%
About one month	9%	10%	8%
A few months	33%	27%	39%
A year or more	27%	31%	24%
Military status of the offender(s)			
Military service member	90%	89%	92%
Higher rank	66%	61%	71%
Similar rank	24%	26%	22%
Lower rank	10%	12%	7%
DoD civilian employee or contractor	9%	10%	7%
Neither or don't know	1%	2%	1%
Work role of the offender(s)			
Supervisor or unit leader	56%	53%	60%
Peer at about the same level	35%	36%	33%
Subordinate	7%	7%	6%
Other	2%	4%	1%
Locations where the behavior ever occurred			
On a military installation/ship	91%	92%	90%
While respondent was on TDY/TAD, at sea, or during field exercises/alerts	25%	27%	23%
While respondent was deployed to a combat zone or to an area where respondent drew imminent danger pay or hostile fire pay	4%	6%	2%
During recruit training/basic training	6%	9%	3%
In a civilian location	28%	24%	32%

Tables B.8.a–B.14.c, for all characteristics). Offenders were most often men, but not always. Among female targets, the offender was a man or men for 93 percent of respondents. Among male targets, this percentage was lower (70 percent). Many offenders continued to sexually harass or discriminate against the target for a long time. One-third (33 percent) of respondents who had experienced sexual harassment or gender discrimination indicated that the situation continued for "a few months" and an additional 27 percent indicated that it continued for "a year or more." One-quarter of targets (23 percent) indicated that it was a single event.

Offenders were typically military service members; 90 percent of targets indicated that the person(s) who sexually harassed or discriminated against them was a military member (or that the group of persons who did it included a military service member). The remaining offenders were either contractors or civilian employees (9 percent) or a non-categorized other (1 percent). Among the 90 percent of offenders who were military, 66 percent were a higher rank than the target (or if it was committed by a group, the group included at least one member of higher rank), 24 percent were "about the same rank," and 10 percent were "of lower rank." Offenders were often the target's supervisor or unit leader; 56 percent of targets indicated that the person who harassed or discriminated against them was their supervisor or unit leader (or that the group targeting them included their supervisor or unit leader). For almost all targets, the harassment or discrimination had occurred on a military installation or ship (91 percent). In sum, the survey data indicate that the sexual harassment and gender discrimination that occurs within the Coast Guard largely involved service member against service member violations, as would be expected given the focus on inappropriate behaviors from "someone at work." Very often the situation reflected a misuse of power by people of higher rank or in a supervisory role to the target.

Effect on Workplace Productivity, the Unit's Mission, and Military Retention

Many targets of sexual harassment and gender discrimination perceived an adverse influence of these negative workplace events on productivity and other workplace-relevant outcomes. The more common perceived workplace outcomes among targets of sexual harassment or gender discrimination were that it caused arguments in the workplace or damaged workplace cohesion (50 percent), made it difficult to complete their work (52 percent), or made the workplace either less productive or compromised the unit's mission (46 percent). Fourteen percent of targets took at least one sick day or other type of leave as a result of the harassment or discrimination, and 25 percent believed that it negatively affected their work evaluations or promotion. See the Annex to Volume 3, Tables B.15.a–B.15.c, for a complete description of targets' perceptions of workplace consequences by gender, service branch, and pay grade.

Sexual harassment and gender discrimination are significant concerns to the Coast Guard not only due to the harm to individuals, but also due to the potential negative effect of these events on the retention of qualified and well-trained service members. Two out of every five Coast Guard members who had been sexually harassed or discriminated against in the past year said that these events had made them "want to leave the military" (39 percent).

In a separate section of the questionnaire, we asked all Coast Guard members whether they were likely to choose to remain on active duty (assuming they had this decision to make). There were notable differences between those who had experienced sexual harassment or gender discrimination in the past year relative to those who had not (Table 4.9). For example, among women who had not been targeted, 7 percent indicated that it was "very unlikely" that they would choose to stay on active duty. Among women who had experienced sexual harassment or gender discrimination in the past year, this percentage increased significantly to 23 percent. For men, the pattern of results is similar, but not statistically significant.

While these empirical findings showing harmful effects of sexual harassment in the Coast Guard are new, these effects were anticipated by the authors of the *Coast Guard Civil Rights Manual* (U.S. Coast Guard, 2010). In particular, the Coast Guard legal definition of sexual harassment includes the following comment:

> The economic costs of sexual harassment are significant. Even more harmful, however, are the negative effects of sexual harassment on productivity and readiness, including increased absenteeism, greater personnel turnover, lower morale, decreased effectiveness, and loss of personal, organizational, and public trust. While not easily quantified, these costs are real and seriously affect the ability of the Coast Guard to accomplish its mission. (p. 2-C.9)

The current study helps to empirically quantify these harmful effects; however, a longitudinal study of service members' responses to sexual harassment and discrimina-

Table 4.9
Self-Reported Likelihood of Choosing to Stay on Active Duty Among Coast Guard Members Who Had Experienced Either Sexual Harassment or Gender Discrimination in the Past Year

Self-Reported Likelihood of Choosing to Stay Active Duty	Sexual Harassment or Gender Discrimination (Men/Women)	None (Men/Women)
Very likely	37% / 26%	55% / 47%
Likely	28% / 19%	22% / 24%
Neither likely nor unlikely	11% / 18%	11% / 12%
Unlikely	11% / 14%	7% / 11%
Very unlikely	13% / 23%	6% / 7%

tion would be a helpful adjunct to these data to better estimate the causal impact of these events on military retention.

Disclosure and Official Reports of Sexual Harassment or Gender Discrimination

One-fifth (20 percent) of targets of harassment or discrimination chose not to tell anyone about their experiences. Men (29 percent) were more likely to keep the situation entirely to themselves than were women (10 percent), and senior officers (2 percent) were less likely than those in other pay grades (20 percent) to choose not to disclose the events to anyone (see the Annex to Volume 3, Tables B16.a–B.16.c, for complete details). Thirty-eight percent of targets disclosed the events only to friends, family, a chaplain, counselor, or medical person (i.e., only to those not formally tasked with investigating or responding to the events).

We identified three types of personnel who are formally required to intervene to stop sexual harassment or gender discrimination when notified of the problem: a work supervisor, someone up the chain of command, or anyone tasked with enforcing MEO regulations. In the sections that follow, we refer to notifying one of these classes of people as "reporting sexual harassment or gender discrimination." We recognize that many of these "reports" can be appropriately handled without generating any official documentation of an allegation of sexual harassment or discrimination. Overall, almost one-half of targets (43 percent) officially reported the violation to someone with the authority and obligation to respond.

Among targets who had reported the events to someone with a formal obligation to respond, we assessed a variety of responses that may have been implemented by actors in the system. Tables B.17a–B.17c in the Annex to Volume 3 provide further detail about these outcomes. Many respondents described responses to their disclosure that are consistent with appropriate and allowable responses for military supervisors, unit leaders, and those tasked with enforcing MEO regulations. These included responses such as someone explaining the rules about sexual harassment to everyone in the workplace (60 percent) and someone speaking with the offender(s) to ask them to change their behavior (54 percent).

However, it was also common for targets to report responses to their disclosure that suggested that the leader had failed to fulfill his or her obligation to respond to MEO complaints. Forty-two percent said the person to whom they reported the event(s) took no action (despite being in a work role that requires the person to take action to address the underlying problem). In some cases, leaders may have taken actions the target was unaware of. On the other hand, about one-third of targets indicated that they had been encouraged to drop the issue (32 percent) or were discouraged from filing an official MEO report (29 percent).

Twenty-eight percent of targets said that the offender(s) retaliated against them for complaining. In fact, a considerable minority of targets also reported experiencing retaliation from coworkers (24 percent) or their supervisor (21 percent).

All survey respondents who experienced sexual harassment or gender discrimination and officially reported the experience were asked about their satisfaction with a variety of aspects of the report or complaint (see Table 4.10). They were asked to respond on a 1–5 scale that ranged from "very dissatisfied" to "very satisfied." Average satisfaction scores across items were around 3, indicating that respondents were typically neither dissatisfied nor satisfied with the response to their report. Tables B.18.a–B.18.c in the Annex to Volume 3 provide additional information about targets' satisfaction with the response to their report by gender, service, and pay grade.

Barriers to Reporting Sexual Harassment and Gender Discrimination
As noted previously, 62 percent of men and 52 percent of women who experienced sexual harassment or gender discrimination in the past year did not report the violation(s) to someone with the authority to respond. For Coast Guard members who did not bring the harassment or discrimination to the attention of someone with the authority to respond, we asked them about their reasons for not doing so. Among the most common reasons for not reporting the harassment or discrimination were that the target thought it was not serious enough to report (58 percent), they wanted to forget about it and move on (46 percent), and that they took other actions to handle the situation (44 percent). Only one significant gender difference emerged; male targets (13 percent) were more likely than female targets (1 percent) to indicate that they did not report because they did not want others to think they were homosexual. See Table 4.11 and Tables B.19.a–B.19.c in the Annex to Volume 3 for additional details.

Table 4.10
Satisfaction with Response to Report of Sexual Harassment or Gender Discrimination

How satisfied were/are you with the following aspects of how the discussion or report was handled?	Mean (SE)
Availability of information about how to file a complaint	3.3 (0.15)
How you were treated by personnel handling your situation	3.0 (0.15)
The action taken by the personnel handling your situation	2.9 (0.14)
The current status of the situation	2.8 (0.13)
Amount of time it took to address your situation	2.8 (0.14)
Availability of information or updates on the status of your report or complaint	2.9 (0.14)

NOTE: Mean based on response scale on which 1 means "very dissatisfied" and 5 means "very satisfied."

Table 4.11
Barriers to Reporting Sexual Harassment and Gender Discrimination

	Total	Men	Women
Minimizing event			
You thought it was not serious enough to report.	58%	65%	48%
You thought a supervisor would make too big of a deal out of it.	34%	33%	35%
You felt partially to blame.	6%	3%	10%
Worried about retaliation			
You thought you might be labeled as a troublemaker.	25%	21%	29%
You were worried about retaliation by the person(s) who did it.	32%	30%	34%
You thought it might hurt your career.	26%	22%	32%
You were worried about retaliation by supervisor or someone in your chain of command.	22%	22%	22%
You thought it might hurt your performance evaluation/ fitness report.	22%	18%	28%
You were worried about retaliation by your military coworkers or peers.	17%	13%	22%
You thought you might get in trouble for something you did.	14%	14%	14%
Concerns about perception			
You did not want people to see you as weak.	31%	29%	33%
You did not want more people to know.	21%	15%	28%
You thought other people would blame you.	17%	13%	21%
You did not want people to think you were gay/lesbian/ bisexual/transgender.	8%	13%	1%
You handled it another way or it didn't need to be handled			
You took other actions to handle the situation.	44%	48%	39%
The offensive behavior stopped on its own.	40%	43%	36%
Someone else already reported it.	3%	3%	3%
Concerns about process			
You did not think anything would be done.	38%	36%	41%
You did not trust the process would be fair.	26%	23%	31%
You did not think you would be believed.	11%	9%	15%
Other			
You wanted to forget about it and move on.	46%	40%	54%
You did not want to hurt the person's career or family.	27%	24%	31%
You did not know how to report it.	3%	3%	3%
Someone told you not to report it.	1%	1%	1%

NOTE: Respondents selected all relevant barriers; therefore, percentages may sum to >100 percent.

Summary

We estimate that 23 percent of active-component Coast Guard women and 5 percent of men experienced sexual harassment or gender discrimination in the past year. Nearly all of the events described by members were events over which the Coast Guard has jurisdiction, and very often, the situation reflected a misuse of power by people of higher rank or in a supervisory role to the target. Consistent with the claims of the *Coast Guard Civil Rights Manual*, we find that productivity, unit cohesion, and retention may be damaged by these violations of professionalism in the workplace. Not all targets chose to report the events to someone with the authority and obligation to respond, but among those who did, the responses were varied. Some targets had outcomes that are consistent with appropriate and allowable responses for military leaders (e.g., someone talked to the person who did it to ask them to change their behavior) whereas others had outcomes that may not be consistent with the leader's obligation to respond (e.g., the target was encouraged to drop the issue or no action was taken). In the latter case, military leaders may have concluded that no violations occurred. Significant barriers to reporting remain in place, including minimization of the event, worries about retaliation, and concern about being stigmatized for reporting. Although the Coast Guard has been taking steps to reduce the rate of these events and mitigate the negative outcomes for targets who choose to come forward, the results of this survey suggest that there remains room for improvement.

Beliefs About Sexual Assault and Sexual Harassment Prevalence, Prevention, and Progress

Kristie L. Gore and Kayla M. Williams

The long form of the 2014 RMWS assessed beliefs and attitudes about safety, prevalence of sexual assault and sexual harassment, reporting, unit leadership, sexual assault prevention training, and expectations for justice following a sexual assault or sexual harassment. What follows is a description of the reported beliefs and attitudes held by different groups within the Coast Guard. Additional descriptive details can be found in Part C of the Annex to Volume 3.

Perceptions of Safety

Most Coast Guard members reported feeling "safe" or "very safe" from being sexually assaulted at their home duty station (92 percent of women and 99 percent of men; Table 5.1). See the Annex to Volume 3, Tables C.1.a–C.2.c, for additional details.

Table 5.1
Perceptions of Safety at Home Duty Station, Estimated Percentages by Gender

	Total	Men	Women
Very safe	86.34% (84.76–87.81)	89.66% (87.82–91.31)	67.52% (64.81–70.14)
Safe	11.24% (9.84–12.76)	8.96% (7.39–10.74)	24.15% (21.78–26.64)
Neither safe nor unsafe	2.02% (1.55–2.58)	1.10% (0.66–1.72)	7.21% (5.80–8.84)
Unsafe	0.10% (0.02–0.30)	NR (NR)	0.70% (0.32–1.33)
Very unsafe	0.30% (0.12–0.61)	0.28% (0.09–0.65)	0.43% (0.13–1.03)

"To what extent do you feel safe from being sexually assaulted at your home duty station."
NOTE: 95-percent confidence intervals for each estimate are indicated in parentheses.

Perceptions of Frequency of Sexual Harassment and Discrimination Against Women

In the Coast Guard, 71 percent of women and 39 percent of men indicated that sexual harassment in the military is either "common" or "very common" (Table 5.2). Sixty-two percent of women and 27 percent of men indicated that discrimination against women is "common" or "very common" in the military (see the Annex to Volume 3, Table C.4.a). A greater proportion of Coast Guard members reported that sexual harassment is "rare" compared with the ratings from members of the Army and Marine Corps; and a lower proportion of Coast Guard members reported that it is "very common" compared to members of the Army, Navy, and Marine Corps (Table 5.3). See the Annex to Volume 3, Tables C.3.a–C.4.c, for additional details.

Table 5.2
Perceptions of Frequency of Sexual Harassment in the Military, Estimated Percentages by Gender

	Total	Men	Women
Very common	7.65% (6.61–8.80)	5.74% (4.61–7.06)	18.40% (16.20–20.76)
Common	35.97% (33.85–38.13)	33.10% (30.66–35.60)	52.15% (49.30–54.99)
Rare	43.79% (41.61–45.99)	46.88% (44.35–49.43)	26.37% (23.96–28.88)
Very rare	12.59% (11.15–14.15)	14.28% (12.59–16.10)	3.08% (2.04–4.47)

"How common is sexual harassment in the military?"
NOTE: 95-percent confidence intervals for each estimate are included in parentheses.

Table 5.3
Perceptions of Frequency of Sexual Harassment in the Military, Estimated Percentages by Service

	Total DoD	Army	Navy	Air Force	Marine Corps	Coast Guard
Very common	11.58% (10.64–12.56)	13.90% (12.27–15.66)	10.86% (8.98–12.99)	7.83% (7.10–8.62)	12.65% (9.48–16.41)	7.65% (6.61–8.80)
Common	38.00% (36.59–39.42)	41.47% (39.17–43.81)	38.06% (34.70–41.51)	33.29% (31.78–34.81)	36.25% (31.58–41.11)	35.97% (33.85–38.13)
Rare	37.66% (36.20–39.13)	33.12% (31.00–35.29)	40.87% (36.92–44.90)	42.32% (40.73–43.91)	36.87% (32.55–41.35)	43.79% (41.61–45.99)
Very rare	12.77% (11.79–13.80)	11.51% (9.73–13.49)	10.21% (8.72–11.86)	16.56% (15.30–17.88)	14.24% (10.87–18.18)	12.59% (11.15–14.15)

"How common is sexual harassment in the military?"
NOTE: 95-percent confidence intervals for each estimate are included in parentheses.

Attitudes and Expectations for Justice

Men are more optimistic than women that sexual harassment and sexual assault will be reported and investigated, and that perpetrators of sexual assault will be held accountable (see the Annex to Volume 3, Tables C.5.a–C.9.c). The Coast Guard ratings on these items are similar to DoD overall, if slightly more confident that justice occurs as it should.

Likelihood of Reporting Behaviors and Taking Action

The large majority of men and women (about 96 percent) indicated they were "likely" or "very likely" to encourage someone who experienced sexual assault to seek counseling. That may be a data point that should be publicized. It reflects a general acceptance toward counseling (or a lack of stigma), and if people were aware that this is a widely held belief, perhaps stigma could be less of a barrier to seeking care for those who need care (see the Annex to Volume 3, Table C.10.a).

Most Coast Guard members indicated that they would be "likely" or "very likely" to encourage someone to report sexual harassment and sexual assault. However, a small percentage of people indicated that they would be "unlikely" or "very unlikely" to report sexual harassment (5 percent of men; 10 percent of women) or sexual assault (3 percent of men; 7 percent of women) if it happened to them. This pattern is consistent with the DoD active-component findings, with a greater proportion of people saying they were "likely" to encourage others to report than to report if it happened to themselves (see the Annex to Volume 3, Table C.10.b and Table C.10.c, for details by pay grade).

Perceptions of Unit Leadership

Women offered lower ratings of unit leadership than men, although most Coast Guard members reported positive views of their leaders' efforts to create an environment free of sexual harassment. About 52 percent of women and 72 percent of men in the Coast Guard indicated "very well" on how their unit leadership "creates an environment where victims would feel comfortable reporting sexual harassment or assault." On average, members of the Coast Guard tended to rate leaders slightly better on these items compared with Army, Navy, and Marine Corps respondents (see the Annex to Volume 3, Tables C.13.a–C.13.c).

Beliefs About Personal Responsibility for Others and Trust in the Military System

Less than 4 percent of Coast Guard members (2.5 percent of men and 8.1 percent of women) reported observing a situation they believed was or could have led to a sexual assault in the past 12 months (see the Annex to Volume 3, Tables C.11.a–C.11.c). This rate is lower than for the other services. Most people intervened in some way (see the Annex to Volume 3, Tables C.12.a–C.12.c). Almost no one in the Coast Guard reported doing nothing in the face of a sexual assault or situation that could lead to one (0.4 percent in the Coast Guard versus 8.7 percent in the DoD active component).

Overall trust in the system was strong in the Coast Guard, with men being slightly more trustful that the military system would treat them respectfully and protect their privacy following a report of sexual assault. Coast Guard beliefs were similar to DoD beliefs on the same items (see Annex to Volume 3, Tables C.17.a–C.17.c).

Perceptions of Progress

A higher proportion of Coast Guard men (20 percent) than women (10 percent) indicated that they believed that sexual assault is "less of a problem" in our nation today than it was two years ago. These Coast Guard perceptions were comparable to those in the DoD military services. Similarly, a higher percentage of Coast Guard men (32 percent) than women (20 percent) reported that sexual assault has become "less of a problem" in the military today than it was two years ago (see the Annex to Volume 3, Tables C.18.a–C.18.c).

Perceptions of and Satisfaction with Sexual Assault Prevention and Response Training

Ninety-nine percent of Coast Guard men and women reported they had training related to sexual assault in the past 12 months. Most "agree" or "strongly agree" that the training covered key topics. Men and women's ratings were similar. Senior officers were somewhat more likely to "strongly agree" that the training covered key topics than junior enlisted Coast Guard members (see Annex to Volume 3, Tables C.14.a–C.15.c). Ninety-eight percent reported they had some training related to sexual harassment in the past 12 months (see Annex to Volume 3, Tables C.16.a–C.16.c).

Conclusion

Generally, beliefs about and attitudes toward risks for sexual assault and sexual harassment were consistent with actual risk. For example, women reported feeling less safe than men and active-component members of the Coast Guard reported greater perceived safety on average than active-component members of the DoD military services. Those at greatest risk for sexual harassment and gender discrimination viewed them as more common than those with lower risk.

CHAPTER SIX

Branch of Service Differences on Measures of Sexual Assault and Sexual Harassment

Terry L. Schell and Andrew R. Morral

Service differences in rates of sexual assault and sexual harassment followed broadly similar patterns for active-component men and women. Specifically, Coast Guard and Air Force personnel experienced lower rates of past-year sexual assault than members of each of the other DoD services. These differences are statistically significant, and some are descriptively large. For instance, Army, Navy, and Marine Corps men are between 3.3 and 5.1 times as likely to have experienced sexual assault in the past year relative to Coast Guard men. Similarly, Army, Navy, and Marine Corps women are between 1.6 and 2.7 times as likely to have experienced a past-year sexual assault relative to Coast Guard women. The rate of past-year sexual harassment of men was also lower than in the DoD services, though the rate of past-year sexual harassment of women in the Coast Guard was similar to the rate among all active-component women in the DoD services.

The magnitude of these differences raises questions about the characteristics of each service that can explain their substantially differing rates of sexual assault and harassment. In this chapter, we explore the possible influence of three types of service differences in explaining the differing risk for sexual assault and harassment. We refer to these classes as *demographic* factors, *military experience* factors, and *military environment* factors. The primary purpose of this analysis is to assess whether demographic differences or differences in deployment experiences account for service differences in sexual assault and harassment risk. Military leaders and policymakers have raised these factors as possible explanations of service differences. In addition, we include several factors, referred to as *military environment* factors, that we know to be associated with risk for these outcomes, based on either our prior statistical analyses (deriving the RMWS sampling weights) or the scientific literature.

- Demographic factors such as age, gender, marital status, ethnicity, qualification test scores, and education level are all associated with sexual assault risk in the military population. To the extent that members of each service differ on these characteristics, this could drive observed differences in risk across services.
- If demographic characteristics—most of which are determined before members join the service—cannot explain service differences in risk, we next consider dif-

ferences between members conferred on them by the military. For instance, the military assigns people to different pay grades, it deploys people to combat zones, and retains them in the military for varying lengths of time.

- If neither the demographic nor the military experience factors explain differences in service risks, we consider a range of military environment variables found to be correlated with sexual assault or harassment risk. These factors include the size of the facility to which the member is assigned and the proportions of the members' unit, installation, and occupational group that are men.

There are, of course, many other differences between services that might be associated with differences in sexual assault risk. There may, for instance, be cultural, policy, or training differences associated with risk. Services could differ in their tolerance of harassment or abuse, in the rigor with which they prosecute offensive or abusive conduct, or in the effectiveness of their sexual assault and sexual harassment training programs. In each case, we might expect such differences to result in service differences in prevalence of sexual assault and sexual harassment. In this chapter, however, we consider only those factors made available to us through the Defense Manpower Data Center's (DMDC's) administrative data.

To evaluate the possible influence of these factors, we conducted a series of analyses on our large active-component sample designed to evaluate the extent that the observed differences among services in the prevalence of sexual assault or harassment could be explained by the demographic characteristics, military characteristics, or military environment differences across services.

We have demographic and military characteristics from DMDC records capturing most such factors known to be associated with sexual assault or harassment. This includes all of the major demographic risk factors for sexual assault that have been identified in prior research on civilian and military samples, with the exception of sexual orientation. We also have measures of the military environment derived from the characteristics of other service members in the same occupational codes, assigned units, and assigned military installations. These environment variables were found to be associated with risk in earlier statistical models and have been identified in the scientific literature as risk factors for sexual assault or harassment. However, we have no individual-level administrative data that capture cultural or policy differences between services. Data on cultural and policy differences would be valuable in future analyses of service differences. Table 6.1 describes the factors derived from DMDC administrative data that were included in our models.

To evaluate the effects of these variables on observed service differences, we modeled the risk ratios for sexual assault and sexual harassment for each service in comparison with the Coast Guard (Table 6.2). *Risk ratios* describe the ratio of the proportion of one group having some experience (such as a past-year sexual assault) to that of another. For instance, the proportion of women in the Marine Corps who experienced

Table 6.1
Factors Considered as Possibly Explaining Service Differences in the Rate of Sexual Assault and Sexual Harassment

Factors	Description
Demographic	
Gender	Men versus women
Age	Age in years
Entry age	Age when joined service
Race	Indicators for Black, White, Hispanic, Asian, Other
Single	Indicator for single versus married
Education	Indicators for four levels of education: high school diploma or less, college without baccalaureate degree, baccalaureate degree, advanced degree
AFQT	Armed Forces Qualification Test score (enlisted only)
Dependents	Number of dependents
Military experiences	
Months deployed (since 7/1/13)	Months of hazardous duty pay in the prior year.
Months deployed (since 9/11/01)	Months of hazardous duty pay during career since 9/11/2001
Pay grade	Seven pay grade categories: E1–E3, E4, E5–E6, E7–E9, W1–W5, O1–O3, O4–O6
AFMS	Career active federal military service (in months)
Military environment	
Occupation male (%)	The proportion of respondent's DoD occupational group who are men
Installation male (%)	The proportion of respondent's assigned installation/ship who are men
Unit male (%)	The proportion of respondent's assigned unit who are men
Installation size	The number of active-duty members assigned to respondents' installation/ship

a sexual assault in the past year was, according to our RMWS results, about 0.0786. The proportion of women in the Coast Guard that had such an experience was 0.0297. Therefore, the unadjusted risk ratio (0.0786/0.0297) is about 2.65, which can be interpreted as indicating that women in the Marine Corps had 2.65 times the risk of experiencing a sexual assault in the past year as did women in the Coast Guard.

Choosing the Coast Guard as the comparison group has no effect on which risk ratios are significantly different from one another. Any service branch could serve as

Table 6.2
Adjusted and Unadjusted Risk Ratios for Sexual Assault Relative to Coast Guard Personnel, by Gender and Service

Gender	Service	Unadjusted Risk Ratio Model 1	Adjusted Risk Ratio Model 2: Demographics	Adjusted Risk Ratio Model 3: Demographics, Mil. Experience	Adjusted Risk Ratio Model 4: Demographics, Mil. Experience, Mil. Environment
Women					
	Coast Guard	1	1	1	1
	Army	1.58 (1.21–2.07)	2.02 (1.54–2.66)	2.04 (1.54–2.70)	2.22 (1.66–2.97)
	Navy	2.18 (1.65–2.88)	2.03 (1.54–2.68)	2.03 (1.54–2.69)	2.20 (1.65–2.92)
	Air Force	0.98 (0.75–1.28)	1.11 (0.85–1.45)	1.12 (0.85–1.46)	1.25 (0.95–1.65)
	Marine Corps	2.65 (1.96–3.58)	2.27 (1.68–3.07)	2.29 (1.69–3.11)	2.14 (1.56–2.94)
Men					
	Coast Guard	1	1	1	1
	Army	3.25 (1.25–8.48)	2.88 (1.10–7.56)	3.60 (1.31–9.92)	4.13 (1.48–11.54)
	Navy	5.09 (1.89–13.71)	4.05 (1.50–10.89)	4.75 (1.76–12.81)	5.06 (1.85–13.80)
	Air Force	1.00 (0.38–2.64)	0.85 (0.32–2.26)	0.97 (0.36–2.63)	0.99 (0.37–2.65)
	Marine Corps	3.89 (1.37–11.08)	2.93 (1.06–8.07)	3.44 (1.23–9.62)	4.25 (1.51–11.99)

NOTE: The risk ratio is the risk of sexual assault in each service relative to the risk to Coast Guard personnel. 95-percent confidence intervals are included in parentheses.

the comparison group and the model results would produce a pattern of significant differences equivalent to those in Table 6.2.

In addition to producing unadjusted risk ratios, the model estimates an adjusted risk ratio that controls for the association of covariates with the outcome.[1] To the

[1] The specific model used to estimate these effects employed a log link-function, so that exponentiated model coefficients were risk ratios rather than odds ratios, as would be produced in a logistic regression model. The models used robust standard errors (i.e., General Estimating Equations), rather than inferring statistical significance directly from a Poisson distribution. All models were estimated using RMWS weights within SAS PROC GENMOD. Models were stratified by gender, thus they always control for a gender effect (even in the unadjusted estimates) as well as all interactions by gender.

extent that differences in the risk for sexual assault between the Coast Guard and other services can be explained by the variables in the model, their risk ratios would move toward 1 in these models. For example, if the risk ratio for women in the Marine Corps relative to the Air Force goes from 2.65 (unadjusted) to 1.00 adjusting for demographic factors, this implies that the differences in prevalence across those services can be fully explained by demographic differences between the Coast Guard and the Marine Corps. In contrast, if the risk ratio grows larger when controlling for demographic factors, it would indicate that the Coast Guard rates were low in spite of (rather than because of) the demographic characteristics that put them at risk.

The three classes of covariates are entered in a specific order. The first adjustment is for demographic factors that largely pre-date a service member's military service or are outside the direct control of the services. The second adjustment adds military experience covariates to the demographic factors; the military experience factors relate to the branch of services' personnel structure and mission. The final adjustment adds to the covariates measures that assess the military environment, which is primarily determined by the gender balance (or gender segregation) of the members' occupation, unit, and installation. This is entered separately from military experience variables largely because these factors may be the result of service policies regarding the integration of women, and thus may be more directly under a service's control.

The column labeled Model 1 in Table 6.2 displays each service's unadjusted risk ratio for sexual assault in comparison with the Coast Guard. With the exception of the Air Force risk ratios, each of these rates for men and women is significantly greater than a risk ratio of 1, indicating higher risk for sexual assault for both men and women in those services relative to the Coast Guard. This can be seen in the 95-percent confidence intervals for Army, Navy, and Marine Corps estimates, which do not include 1.00.

Model 2 provides risk ratios comparing each service with the Coast Guard while adjusting for demographic characteristics. With the exception of the Air Force, the risk

---

In addition to the predictors listed in Table 6.1, the regression models included a range of additional terms. These include: (1) missing data flags for cases that were missing *Entry Age*, *Education*, *AFQT*, and *assigned unit* to avoid case-wise deletion on covariates with nontrivial missingness; and (2) quadratic terms for the effects of *Age* and *AFMS*. The model results presented in Table 6.2 do not include two-way interactions between all covariates, but we did explore whether inclusion of interaction terms would change the pattern of results. Specifically, because of the large number of two-way interactions, we explored adding interaction terms to the base model (the model with main effects, missing variable flags, and quadratic terms) for every term that was significant at the $p < 0.15$ level in the final model (Model 4). However, if an interaction met this entry criterion it was included in Models 2 or 3 when it was between two variables that were also included in those models. The only exception to these rules was Model 4, predicting risk for sexual assault among men. The small number of assaulted men relative to the number of predictors in the model resulted in estimation problems; for this one model, the main effects of variables listed in Table 6.1 were also removed from the model if they were not significant at $p < 0.15$ to create a more parsimonious model. The results of models that included interaction terms were nearly identical to the model without them that we present in Table 6.2. Specifically, no risk ratio differed by more than 0.31 from the base model, and the pattern of significance was identical.

ratios all remain significantly greater than 1 (the rate in the Coast Guard). However, the differences among the Army, Navy, and Marine Corps are reduced in comparison to the unadjusted risk ratios. That is, the lower sexual assault risk for the Coast Guard relative to the Army, Navy, and Marine Corps is not fully explained by demographic differences across services. Interestingly, however, demographic differences do seem to explain the significant differences among the Army, Navy, and Marine Corps. It appears, therefore, that the apparent differences in risk between services are partially explained by demographic factors, with the exception of the low rates in the Coast Guard and Air Force.

Model 3 adds military experience variables to the demographic factors. However, these variables appear to affect the risk ratios only minimally while controlling for the demographic characteristics, and do not explain the differences between each service and the Coast Guard.

Finally, Model 4 adds military environment factors to all the previously included variables, and differences in risk of past-year sexual assault remain. Risk for Coast Guard personnel remains significantly lower than that found in the Army, Navy, and Marine Corps for both men and women. It is nearly equivalent to the adjusted risk found in the Air Force. In contrast, the differences in sexual assault risk among the Army, Navy, and Marine Corps are almost fully explained by the variables in Model 4. The remaining service differences relative to the Coast Guard and Air Force are descriptively large; men in those three services are 4 to 5 times as likely to experience a sexual assault as are Coast Guard men with comparable demographic characteristics, military experiences, and military environments.

Table 6.3 presents comparable analyses of risk ratios for experiences of sexual harassment in the past year.[2] In the unadjusted Model 1 results, men and women in the Army, Navy, and Marine Corps are all at significantly higher risk for sexual harassment in the past year than are members of the Coast Guard. Air Force women, in contrast, have significantly lower risk of sexual harassment than women in the Coast Guard, while there is no statistically significant difference between risk for the Air Force men and the Coast Guard men.

Adjustment for our demographic variables in Model 2 causes the Marine Corps risk ratio for men to no longer be significantly greater than the Coast Guard, but otherwise the pattern of significant differences remains unchanged from Model 1. Controlling for demographic characteristics, women in the Army, Navy, and Marine Corps are about 1.4 times as likely to experience sexual harassment in the past year as women in

---

[2] As with our analyses of service differences in sexual assault, we explored whether inclusion of any two-way interactions among model covariates would substantively change the findings. Using the same procedures as with sexual assault, the model that included interaction terms looked nearly identical to that excluding those terms. Indeed, no risk ratio in the model with interaction terms differed by more than 0.06 from those in Table 6.3, and the patterns of significance are identical between the base model and the model with interaction terms.

Table 6.3
Adjusted and Unadjusted Risk Ratios for Sexual Harassment Relative to Coast Guard Personnel, by Gender and Service

Gender	Service	Unadjusted Risk Ratio Model 1	Adjusted Risk Ratio Model 2: Demographics	Adjusted Risk Ratio Model 3: Demographics, Mil. Experience	Adjusted Risk Ratio Model 4: Demographics, Mil. Experience, Mil. Environment
Women					
	Coast Guard	1	1	1	1
	Army	1.20 (1.07–1.35)	1.46 (1.30–1.65)	1.44 (1.28–1.63)	1.54 (1.35–1.75)
	Navy	1.45 (1.28–1.64)	1.46 (1.29–1.65)	1.46 (1.29–1.65)	1.53 (1.35–1.73)
	Air Force	0.65 (0.57–0.73)	0.72 (0.64–0.82)	0.72 (0.64–0.81)	0.80 (0.70–0.90)
	Marine Corps	1.42 (1.23–1.64)	1.39 (1.20–1.61)	1.41 (1.22–1.63)	1.30 (1.11–1.51)
Men					
	Coast Guard	1	1	1	1
	Army	2.05 (1.59–2.64)	1.97 (1.53–2.55)	2.26 (1.69–3.02)	2.11 (1.56–2.86)
	Navy	2.23 (1.69–2.95)	1.98 (1.50–2.61)	2.22 (1.68–2.92)	2.24 (1.69–2.97)
	Air Force	0.88 (0.67–1.15)	0.83 (0.63–1.09)	0.91 (0.69–1.21)	0.97 (0.73–1.30)
	Marine Corps	1.63 (1.18–2.26)	1.36 (0.98–1.90)	1.58 (1.13–2.21)	1.31 (0.93–1.85)

NOTE: The risk ratio is the risk of sexual assault in each service relative to the risk to Coast Guard personnel. 95-percent confidence intervals are included in parentheses.

the Coast Guard, but are at higher risk than Air Force women. Men in the Army and Navy are about twice as likely as Coast Guard men to have such experiences.

Adding military experience variables to the demographic variables has only small effects on risk ratios, and the pattern of significant differences between services is generally similar between Model 2 and Model 3—with one exception. Similar to the unadjusted risk ratios, Marine Corps men have significantly higher risk of sexual harassment than do Coast Guard men in this model. Therefore, differences between the Coast Guard and the Army, Navy, and Marine Corps are not explained by demographic factors, or, for instance, by differences in service tenure or months of deployment in the past year.

The military environment variables included in Model 4 have only a modest effect on most of the risk ratios. These variables are primarily indicators of how "male" a service member's environment is, based on their occupational group, unit, and installation composition. Because the Marine Corps has the lowest proportion of women among the services, and sexual harassment is more common in predominately male environments, adjusting for these factors has the largest effect on the Marine Corps risk ratios. Controlling for all of these variables, the risk of sexual harassment among men is about twice as high in the Army and Navy as the Coast Guard, while the Air Force and Marine Corps show similar rates to the Coast Guard. Among women, Army, Navy, and Marine Corps women are at 1.3 to 1.5 times the risk of experiencing sexual harassment as Coast Guard women. On the other hand, women in the Air Force are at significantly lower risk of sexual harassment than women in the Coast Guard, at about 0.8 times the risk.

Looking across the sexual assault and sexual harassment analyses presented here suggests that differences between the Coast Guard and the four DoD services are not chiefly due to differences in the demographic characteristics of members in each service, nor to the kinds of military experience and military environment factors we considered. The one exception may be the sexual harassment experiences of Marine Corps men in the past year, who appear to have higher rates than the Coast Guard before accounting for demographic and other differences, but who appear to have comparable rates after these adjustments.

Coast Guard rates of sexual assault for men and women are comparable to those found for the Air Force. This is also true for the past-year sexual harassment experiences of men. But in the case of women, the Air Force has significantly lower sexual harassment rates than the Coast Guard, differences that are not attributable to any of the factors we considered in the present analysis.

CHAPTER SEVEN

Findings from the Coast Guard Reserve

Terry L. Schell and Andrew R. Morral

The RMWS survey included about 2,500 respondents who were Coast Guard reservists. Similar to the prior versions of the Workplace and Gender Relations Survey of Reserve Members, only members of the Selected Reserve were sampled, not members of the Individual Ready Reserve, the Standby Reserve, the Retired Reserve, or the Coast Guard Auxiliary.

The RMWS was not intended to provide a comprehensive assessment of the experiences of Coast Guard Reserve members. Reserve members all received the RMWS short or medium form; therefore, we do not have measures of unwanted sexual contact, perceptions of risk for sexual assault or harassment, or opinions about military climate. As such, we provide top-line comparisons only, between members of the Coast Guard active and reserve components as measured on the RAND form. Additional information about the reserve sample is contained in the appendix.

Sexual Assault

Estimated rates of sexual assault in the past year for Coast Guard Reserve members are presented in Table 7.1.[1] Across the entire sample frame of 7,592 reservists who were below flag officer in rank and who had served for at least six months, these rates correspond to approximately 40 individuals assaulted in the past year, with a 95-percent confidence interval ranging from 20 to 60. Because of the comparatively low prevalence of sexual assault against reserve men relative to women, the majority of all Coast Guard Reserve members who experienced a sexual assault were women, despite the fact that women made up just 17 percent of this population. The rates of sexual assault for men and women in the Coast Guard Reserve was not significantly different from rates estimated among men and women in the DoD reserve component.

Although the past-year sexual assault point estimates for both men and women in the Coast Guard Reserve appear lower than for active-component members (0.2 percent

---

[1] As with active-component members, the survey counts all past-year sexual assaults, not just those that occurred while the reservist was drilling or in an active status.

Table 7.1
Estimated Percentage of Coast Guard Reserve Members Who Experienced a Sexual Assault in the Past Year, by Gender and Assault Type

Grouping	Total	Men	Women
Any sexual assault	0.47% (0.23–0.85)	0.22% (0.07–0.53)	1.71% (0.63–3.71)
Penetrative	0.23% (0.07–0.53)	0.04% (0.00–0.26)	1.15% (0.31–2.93)
Non-penetrative	0.25% (0.09–0.55)	0.18% (0.04–0.47)	0.58% (0.07–2.08)
Attempted penetrative	0.00% (0.00–0.13)	0.00% (0.00–0.16)	0.00% (0.00–0.70)

NOTE: 95-percent confidence intervals are included in parentheses.

versus 0.3 percent for men; 1.7 percent versus 3.0 percent for women), these differences are not statistically significant (see Table 3.2). However, this lack of significance may be partially attributable to a lack of precision in estimating these percentages within the available sample.

Because of the low number of survey respondents who experienced any type of sexual assault in the past year, it is difficult to detect any pattern in the rates of sexual assault across pay grades (Table 7.2). For instance, although the estimated rate for women who were in E1–E4 pay grades was more than three times higher than for those in E5–E9 pay grades, this difference is not statistically significant.

As noted in Chapter Two, our reporting standards protect respondent privacy by withholding any analyses on groups with fewer than 15 members. Because the rates of sexual assault are very low in the Coast Guard Reserve sample, this policy means we are unable to characterize the experiences of those who were assaulted in the past year.

Table 7.2
Estimated Percentage of Coast Guard Reserve Members Who Experienced Any Type of Sexual Assault in the Past Year, by Gender and Pay Grade

Pay Grade	Total	Men	Women
Total	0.47% (0.23–0.85)	0.22% (0.07–0.53)	1.71% (0.63–3.71)
E1–E4	0.59% (0.14–1.60)	0.00% (0–0.65)	3.70% (1.03–9.15)
E5–E9	0.53% (0.21–1.11)	0.43% (0.13–1.04)	1.06% (0.13–3.77)
O1–O6	0.00% (0.00–0.55)	0.00% (0.00–0.71)	0.00% (0.00–2.42)

NOTE: 95-percent confidence intervals are included in parentheses. Pay grade categories are collapsed to improve precision of estimates.

Sexual Harassment and Gender Discrimination

The assessment of sexual harassment and gender discrimination among reserve-component members differed from those administered to active-component members. Reserve-component members were asked about workplace experiences that occurred "while you were on military duty, including National Guard or reserve duty such as weekend drills, annual training, and any period in which you were on active duty. Do not include experiences that happened in your non-military job." That is, they were asked to limit their responses to describing experiences that occurred at their military workplaces, excluding events in their civilian workplace in the past year. In contrast, active-component personnel were simply asked about their workplace experiences. This difference in question wording is important for understanding differences between the active and reserve components on sexual harassment outcomes; reservists spend far less time at their military workplace relative to active-component Coast Guard members.

We estimate that approximately 4 percent of Coast Guard Reserve members experienced sexual harassment or gender discrimination in the past year. However, the risk for such violations varied substantially by gender, with 2 percent of men and 14 percent of women experiencing these violations (Table 7.3). As with the Coast Guard active component, the majority of these violations for both men and women involved sexual harassment of the sexually hostile workplace type.

These rates of sexual harassment and gender discrimination among the Coast Guard Reserve are significantly lower than was found in the Coast Guard active component (see Chapter Four). For example, we estimated that 4 percent of the Coast

Table 7.3
Estimated Percentage of Coast Guard Reserve Members Who Experienced a Sex-Based MEO Violation in the Past Year, by Gender and Type

Grouping	Total	Men	Women
Any sex-based MEO violation	4.24% (3.14–5.58)	2.22% (1.36–3.39)	13.98% (9.55-19.49)
Gender discrimination	1.61% (0.99–2.47)	0.80% (0.35–1.57)	5.52% (2.94–9.33)
Any sexual harassment	3.31% (2.33–4.56)	1.80% (1.02–2.93)	10.59% (6.67–15.74)
Sexually hostile environment	3.31% (2.33–4.55)	1.80% (1.02–2.92)	10.59% (6.67–15.74)
Sexual *quid pro quo*	0.07% (0.00–0.42)	0.00% (0.00–0.24)	0.39% (0.01–2.22)

NOTES: 95-percent confidence intervals are included in parentheses. Any sex-based MEO violation includes past-year experiences of either gender discrimination or sexual harassment. Any sexual harassment includes past-year experiences of either a sexually hostile environment or sexual *quid pro quo*.

Guard Reserve versus 7 percent of the active component experienced any type of MEO violation in the past year. Specifically, when compared to the Coast Guard active component, both male and female reservists had significantly lower rates of sexual harassment. While women in the reserve component were less likely to experience military gender discrimination in the past year than women in the active component, this difference was not statistically significant for men. The differences among women may be due to the fact that reservists spend much less time in their military workplace than active-component service members, so there is less opportunity for a violation to occur.

CHAPTER EIGHT

Discussion and Recommendations

Andrew R. Morral, Kristie Gore, and Terry L. Schell

The 2014 RMWS is one of the largest studies of sexual assault and sexual harassment among members of the military ever conducted, and the first to estimate criminal sexual assault as defined by the UCMJ, as well as sexual harassment and gender discrimination as codified in DoD and Coast Guard regulations. High rates of participation by sampled Coast Guard members resulted in more than 7,000 survey responses from active-component members, including more than one-half of all Coast Guard women, and 2,500 members of the Coast Guard Reserve.

The large sample and high response rates provided the opportunity to conduct detailed investigations into relatively rare events. We are able to estimate, for instance, that of more than 39,000 active-component Coast Guard members, between 180 and 390 experienced a criminal sexual assault in the past year. Nevertheless, some experiences, such as sexual assaults of men in the past year, are sufficiently infrequent in the Coast Guard that we are unable to provide more than top-line estimates of their prevalence. In this final chapter, we summarize some of the key findings from the study and propose steps that should be considered for further reducing the prevalence of sexual assault and sexual harassment against members of the Coast Guard.

Sexual Assault

Approximately 3 percent of women and 0.3 percent of men in the active component of the Coast Guard experienced one or more sexual assaults in the past year. These rates are low in comparison with the Army, Navy, and Marine Corps, but comparable to the Air Force. Indeed, even after accounting for demographic and other differences between members of each service, women in the Army, Navy, and Marine Corps were more than twice as likely to have been sexually assaulted in the past year, and men in those services were four to five times as likely to have been sexually assaulted, compared with women and men in the Coast Guard.

We were able, therefore, to rule out many of the factors that have been proposed to explain why some services have lower rates of sexual assaults than others. In particular, the Coast Guard's low rates were not because its members are older, more likely to be married, more highly educated, have been deployed for fewer months, or any of the other factors considered in Chapter Six. Other differences between the Coast Guard and the three DoD services with higher rates of past-year sexual assault do exist and were not included in our statistical model. We cannot say whether the key explanatory differences are cultural differences between the services, differences in training, differences in patterns of life members experience (such as where they are quartered or the amount of time they spend away from home), or other alternative explanations. However such factors should be investigated in subsequent research so that the Coast Guard can identify and further promote factors that reduce the rates of sexual assault in the military.

Among women in the Coast Guard who were assaulted in the past year, the assailant was another member of the military in 77 percent of all cases. This rate is significantly lower than the proportion of women assaulted by a member of the military across all DoD services (89 percent), although the proportion among sexually assaulted women in the Coast Guard was similar to the Air Force.

When a sexual assault occurred against Coast Guard women, alcohol was more frequently involved than among women in most other services. Indeed, more than 75 percent of assaults against Coast Guard women occurred after either the woman or the assailant had been drinking. In contrast, 56 percent of assaults against women in DoD services occurred after alcohol consumption by the victim or the assailant. This higher proportion of sexual assaults involving alcohol is consistent with other results showing that Coast Guard women were at lower risk of sexual assault at work than women in some other DoD services. For example, assaults against Coast Guard women more commonly occurred while out with friends or at a party.

Sexual Harassment and Gender Discrimination

Far more Coast Guard members experienced sexual harassment in the past year than experienced sexual assault. We estimate that approximately 6 percent of active-component Coast Guard members, or 2,350 members, experienced some form of sexual harassment in the past year. A higher proportion of women (1 out of 5) than men (1 out of 25) had workplace experiences in the past year that under Coast Guard directives would be classified as sexual harassment, an unfair condition of their employment.

That sexual harassment is relatively common within the Coast Guard is widely recognized by service women, 70 percent of whom indicated that sexual harassment in the military is either "common" or "very common." Fewer servicemen hold this view (39 percent). These rates are comparable to those found across DoD services, where 69 percent of women and 34 percent of men describe sexual harassment as common.

Men and women in the Coast Guard were less likely than members of the Army, Navy, or Marine Corps to be sexually harassed in the past year. However, women in the Coast Guard were more likely to be sexually harassed than Air Force women. Moreover, these differences were not fully explained by any of the demographic or other differences between the members of each service. Even after adjusting rates of sexual harassment to account for age, education, marital status, deployment experience, and many other factors, the risk of sexual harassment among men was about twice as high in the Army and Navy as the Coast Guard, while Air Force and Marine Corps men showed similar rates to the Coast Guard. Among women, Army, Navy, and Marine Corps women were at 1.3 to 1.5 times the risk of experiencing sexual harassment as Coast Guard women.

On the other hand, even after adjusting for demographic and other differences, women in the Air Force were at significantly lower risk of sexual harassment than women in the Coast Guard. Indeed, Coast Guard women had 1.25 times the risk of past-year sexual harassment than did Air Force women. Given the many similarities between the Coast Guard and Air Force in terms of risks of sexual assault for men and women, and risk of sexual harassment for men, the fact that sexual harassment risks to women in the Coast Guard were significantly higher than in the Air Force stands out as an anomaly worthy of investigation and correction.

Although less common than sexual harassment, approximately 3 percent (1,020) of Coast Guard active-component members believed their careers were harmed in the past year because of gender discrimination, with women being more than ten times as likely as men to be classified as having such an experience in the past year. Like sexual harassment, gender discrimination against women is widely recognized as an issue for the Coast Guard, at least among women, 62 percent of whom described discrimination against women as common or very common in the military, compared with 27 percent of men.

Men in the Coast Guard were at lower risk of gender discrimination than men in the Army and Navy, and had rates that were not significantly different than those found for men in the Marine Corps or the Air Force. Again, however, rates of gender discrimination for women in the Coast Guard were significantly greater than experienced by women in the Air Force, and significantly lower than those of women in the Army, Navy, and Marine Corps. Thus, despite comparable rates of past-year sexual assault for members of the Coast Guard and Air Force, Coast Guard women appeared to experience significantly greater rates of both gender discrimination and sexual harassment than their peers in the Air Force.

The substantial majority of Coast Guard members who experienced sexual harassment or gender discrimination described their offender(s) as members of the military (90 percent). In two-thirds of the incidents that involved a military service member, one or more of the offenders were of higher rank than the target, and more than one-half of the time the offender or offenders included a supervisor or unit leader.

The sexual harassment and gender discrimination experienced by Coast Guard members was usually an ongoing problem rather than an isolated incident. We estimated that 60 percent of targets were harassed or discriminated against for three months or more.

Overall, about one-half of targets disclosed the sexual harassment or gender discrimination to a work supervisor, a unit leader, or someone tasked with enforcing MEO regulations. In many cases, that person took appropriate steps in response, but there was also evidence that some leaders failed to fulfill their obligations after learning about a violation.

Coast Guard men and women who we classified as experiencing either sexual harassment or gender discrimination in the past year described a range of related workplace harms. Many of these people indicated that the incidents caused arguments in the workplace or damaged workplace cohesion (50 percent), made it difficult for them to complete their work (52 percent), or made the workplace either less productive or compromised the unit's mission (46 percent). Fourteen percent took at least one sick day or other type of leave as a result of the harassment or discrimination. Two out of every five Coast Guard members who had been sexually harassed or discriminated against in the past year indicated that these events had made them "want to leave the military" (39 percent). Among women who had not experienced sexual harassment or gender discrimination, 7 percent indicated that it was "very unlikely" that they would choose to stay active duty. However, among women who had experienced sexual harassment or gender discrimination in the past year, this percentage increased significantly to 23 percent. For men, the pattern of results is similar, but not statistically significant.

Sexual harassment and gender discrimination may also contribute to the risk of sexual assault. Certainly the correlation between the two is strong. Women who experienced sexual harassment in the past year were 14 times more likely to have been classified as experiencing sexual assault during the same period. Moreover, 30 percent of women who were sexually assaulted indicated that their assailant had sexually harassed them prior to the assault.

These empirical findings support what was known or suspected by the authors of the *Coast Guard Civil Rights Manual* (U.S. Coast Guard, 2010). In particular, the Coast Guard legal definition of sexual harassment includes the following comment:

> The economic costs of sexual harassment are significant. Even more harmful, however, are the negative effects of sexual harassment on productivity and readiness, including increased absenteeism, greater personnel turnover, lower morale, decreased effectiveness, and loss of personal, organizational, and public trust. While not easily quantified, these costs are real and seriously affect the ability of the Coast Guard to accomplish its mission. (p. 2-C.9)

The current study helps to empirically quantify these harmful effects; however, a longitudinal study of service members' responses to sexual harassment and gender discrimination would be a helpful adjunct to these data to better estimate the causal impact of these events on military retention.

Recommendations

1. *Concentrate additional prevention and enforcement efforts on sexual harassment and gender discrimination.* Reducing the incidence of sexual harassment and gender discrimination is likely to have far-reaching benefits for the Coast Guard, possibly including improved workplace productivity, reduced sick time, and improved recruitment and retention, and it may reduce the prevalence of sexual assault. Moreover, there are good reasons to suspect that these violations can be reduced. Specifically, the Air Force, which has an annual rate of sexual assault against men and women comparable to that of the Coast Guard, nevertheless has significantly lower rates of both sexual harassment and gender discrimination against women. Although we cannot isolate the factors that explain this difference between the Coast Guard and the Air Force, we can rule out many likely causes, such as the demographic and service attributes we examined in Chapter Six.

2. *Review training and instructional materials to ensure that they make clear that some reportable sexual assaults may occur in the context of hazing or bullying or may not be perceived by either the service member or the offender as a sexual encounter.* Sexual assaults against men are sufficiently infrequent that we could not, with this survey, characterize them beyond offering an overall prevalence estimate. Based on findings from the much larger sample of DoD men (see Volume 2), we suspect that many assaults against men occur as a part of hazing activities or as a form of harassment or bullying. However, some service members may not recognize that unwanted touching of private parts or penetration may qualify as a UCMJ Article 120 sexual assault, even if it were not done for sexual gratification, provided the intent of the contact was to abuse, harass, or humiliate. Ensuring that members of the Coast Guard understand the full scope of events that qualify as sexual assaults may improve reporting and provide those who are being abused with needed response systems.

3. *Develop monitoring systems for sexual harassment, gender discrimination, hazing, bullying, and physical assaults.* Sexual assault is sufficiently infrequent in the Coast Guard that vast numbers of its members must be surveyed to estimate the prevalence. The same is not true for sexual harassment, which was experienced by approximately 6 percent of all members in the past year, including 19 percent of all women. This comparatively high prevalence rate makes it possible to generate estimates of the extent of the problem for smaller samples of respondents, including, for instance, members assigned to individual commands, installations, or possibly ships. For reasons described above, we believe it might be valuable to extend this monitoring to cover not just MEO violations, but hazing, bullying, and other misconduct as well, all of which form a nexus

that may contribute to sexual assault risk and to undermining good order and discipline in the Coast Guard.

4. *Investigate the causes and consequences of sexual assault.* The RMWS has provided unprecedented detail on the nature and circumstances of sexual assault, sexual harassment, and gender discrimination in the military services, but the new insights offered by these data raise new questions that the Coast Guard should consider investigating further. Specifically:

   a. We find significant differences between the risk of sexual assault to which Coast Guard members are exposed in comparison with that for members of the Army, Navy, and Marine Corps. Although we have ruled out many plausible risk factors on which members of each of these services may differ from the Coast Guard, we have not identified any risk factors that can explain the Coast Guard's lower risk. If we were able to determine that risk differences are attributable to cultural differences between the services, differences in training, differences in patterns of life members experience (such as where they are quartered or the amount of time they spend away from home), or other such factors, this could provide important insights into how to further reduce sexual assault risk in the Coast Guard, in other military services, and possibly in civilian settings as well.

   b. Our results raise the possibility that sexual harassment and gender discrimination may have a range of harmful effects on service members' careers, their safety, and their retention in the Coast Guard. A longitudinal study of service members' responses to sexual harassment and gender discrimination would be a helpful adjunct to these data to better estimate the consequences for the Coast Guard of these incidents.

Additional Information on the RAND Military Workplace Study

This report is the third in a series on the RAND Military Workplace Study. Additional information about the study design, the survey instrument, and its rationale can be found in Volume 1. Volume 2 describes findings for the Army, Navy, Air Force, and Marine Corps. Finally, Volume 4 presents a series of methodological investigations each designed to better understand possible sources of bias in our survey results. This includes additional data collection to estimate differences in the sexual assault experiences between those who chose to complete the survey and those who did not; undercounting or overcounting of past-year sexual assaults because of who was included and excluded from the sample frame; biases resulting from counting events as occurring in the past year that actually occurred earlier; or counting events as crimes that were not.

The Coast Guard Sample

Bonnie Ghosh-Dastidar and Terry L. Schell

Active Component

Sample frame. The population included all Coast Guard active-component members listed in the May 2014 Defense Enrollment Eligibility Reporting System (DEERS) database maintained by the DMDC, a population of 39,112. For continuity with earlier WGRA surveys, we matched the exclusion criteria previously used to define WGRA sampling frames (see Volume 1 for details).[1]

Sample selection. The active-component sample included all women in the sample frame and 25 percent of men. The sample sizes were designed to provide enough respondents who had experienced a sexual assault in the past year so that the characteristics of those assaults could be analyzed with sufficient statistical precision. The resulting Coast Guard sample included 14,167 active-component members, of whom 32.7 percent were women. The composition of the sampling frame and the sample for all active-component (including DoD and Coast Guard) service members is listed in Table A.1.

Reserve Component

Sample frame. The population included members of the Coast Guard Reserve listed in the May 2014 DMDC dataset—a population of 7,592. Exclusion criteria are similar to those for the active-component sample (see Volume 1 for details).

---

[1] Those with fewer than six months of service have historically been excluded from WGRA surveys for logistical and substantive reasons. In terms of survey logistics, the development of a sample frame and survey fielding historically have taken several months, so it has not been possible to enter the field pursuing a sample that has fewer than several months of service. In addition, those still in basic training or transitioning to their first assigned units are difficult to reach, as their addresses and even email addresses are likely to have changed between the time the sample is drawn and the field date of the survey. Substantively, those with fewer than six months of service can provide only a partial estimate for the main "past year" measures in the WGRA. Alternative sampling and survey methods would need to be employed to get accurate population estimates of newer service members.

General and flag officers have been excluded in the past (and in the RMWS) because, as the leaders and decisionmakers in the services, their experience is not expected to be comparable to others, and their numbers are too small to satisfactorily analyze separately.

Table A.1
Coast Guard Active-Component Sampling Frame and Sample Sizes, by Gender, Service, and Pay Grade

	Total		Women		Men	
	Population	Sample	Population	Sample	Population	Sample
Total number	1,356,673	491,680	203,343	203,343	1,153,330	288,337
Column percentages:						
Army	37.1%	36.2%	34.2%	34.2%	37.6%	37.6%
E1–E4	15.8%	15.6%	15.2%	15.2%	15.9%	15.9%
E5–E9	15.3%	14.2%	11.9%	11.9%	15.9%	15.9%
O1–O3	3.6%	4.0%	4.8%	4.8%	3.4%	3.4%
O4–O6	2.3%	2.3%	2.3%	2.3%	2.3%	2.3%
Navy	23.1%	24.3%	27.0%	27.0%	22.4%	22.4%
E1–E4	9.3%	10.6%	13.6%	13.6%	8.5%	8.5%
E5–E9	10.0%	9.8%	9.2%	9.2%	10.2%	10.2%
O1–O3	2.3%	2.5%	2.9%	2.9%	2.1%	2.1%
O4–O6	1.5%	1.5%	1.3%	1.3%	1.5%	1.5%
Air Force	23.2%	25.1%	29.2%	29.2%	22.1%	22.1%
E1–E4	8.2%	8.7%	9.9%	9.9%	8.0%	8.0%
E5–E9	10.3%	11.2%	13.2%	13.2%	9.8%	9.8%
O1–O3	2.6%	3.0%	4.0%	4.0%	2.3%	2.3%
O4–O6	2.1%	2.1%	2.1%	2.1%	2.0%	2.0%
Marine Corps	13.7%	11.6%	6.8%	6.8%	15.0%	15.0%
E1–E4	8.1%	6.9%	4.3%	4.3%	8.8%	8.8%
E5–E9	4.2%	3.5%	1.9%	1.9%	4.6%	4.6%
O1–O3	0.9%	0.8%	0.5%	0.5%	1.0%	1.0%
O4–O6	0.5%	0.4%	0.1%	0.1%	0.5%	0.5%
Coast Guard	2.9%	2.9%	2.9%	2.9%	2.9%	2.9%
E1–E4	0.9%	1.0%	1.2%	1.2%	0.8%	0.8%
E5–E9	1.5%	1.3%	1.0%	1.0%	1.6%	1.6%
O1–O3	0.3%	0.3%	0.4%	0.4%	0.3%	0.3%
O4–O6	0.2%	0.2%	0.2%	0.2%	0.2%	0.2%

NOTES: Warrant officers are included in the E5–E9 group for the purposes of sampling.

Sample selection. Due to the small number of members in the Coast Guard Reserve, we included every reserve member in the survey. The reserve-component sample size was smaller than the active-component sample size because we will not produce separate prevalence estimates for detailed reporting categories. The Coast Guard Reserve sample included 1,267 women and 6,325 men.

Final Respondent Disposition

Service members included in the 2014 sample were considered *eligible* if they were alive at the end of the survey field period. Our definition of *eligible complete* included anyone whose sexual assault status could be determined. We classified *eligible nonrespondents* into four groups: no response, active refusal, partial complete with no information, and partial complete with insufficient information. The partial completes were separated into two groups to distinguish between those participants who started the survey and provided no information versus those who provided some but insufficient information to determine whether they were sexually assaulted in the past year (see Volume 1 for further details).

Active Component

Table A.2 summarizes the case disposition categories for the active-component sample, which follow survey research standards for documentation (American Association for Public Opinion Research, 2011).

Table A.2
Case Disposition Frequencies for the Coast Guard Active-Component Sample

Case Disposition	Sample cases	Percentage
Total sample	14,167	100.0
Ineligible—deceased	0	0.0
Eligible complete	7,307	51.6
Nonresponse		
No response	6,240	44.0
Active refusal	27	0.2
Partial complete, no information	312	2.2
Partial complete, insufficient information	281	2.0

NOTE: *Partial complete, no information* refers to sampled members who loaded the survey consent form but did not complete any survey questions. *Partial complete, insufficient information* refers to sampled members who answered at least one survey question, but were missing the measure of sexual assault or unwanted sexual contact.

Of 14,167 sampled records, there were 7,307 eligible completes (51.6 percent). Another 44 percent of the sample provided no response after repeated attempts to reach the service member. Of the partial respondents, 53 percent provided no information while the remainder provided insufficient information to determine whether they had experienced a sexual assault in the past year.

Tables A.3 and A.4 provide information on the quality of the postal and email addresses for the active-component sample. In the first mailing sent to the full sample, less than one percent of addresses were identified in the National Change of Address (NCOA) system as an unmailable address. NCOA processing identifies individuals who have submitted address changes within the past 12 months, in addition to verifying that the mailing address is valid, with a matching city and zip code. Another 10.6 percent were returned as postal non-deliverable. Mail with bad addresses was returned by the postmaster as non-deliverable.

Sample members could have multiple email addresses. The email addresses were ordered by priority on the sample record, with the military email address considered first priority and home email addresses considered second priority. Surveys were programmed to be sent to the highest-priority email address. No email sent indicates that no address was available. In the first batch of emails sent, 3 percent of the sample was missing an email address, while another 4.5 percent encountered a bounce-back due to a non-working email address.

Reserve Component

Table A.5 provides the breakdown by case disposition categories of the Coast Guard Reserve sample. Out of a sample of 7,592 Coast Guard Reserve members, there were 2,537 eligible completes. The percentage of the sample in the reserve component

Table A.3
Quality of Mailing Address Based on Initial Mailing

	Sample cases	Percentage
Total sample	14,167	100.0
No mail sent	83	0.6
Mail non-deliverable	1,507	10.6

Table A.4
Quality of Email Address Based on Initial Email

Case Disposition	Sample cases	Percentage
Total sample	14,167	100.0
No e-mail sent	440	3.1
Bounce back	640	4.5

Table A.5
Case Disposition Frequencies for Coast Guard Reserve Sample

Case Disposition	Sample Cases	Percentage
Total sample	7,592	100.0
Ineligible—deceased	0	0.0
Eligible complete	2,537	33.4
Nonresponse		
No response	4,839	63.7
Active refusal	4	0.0
Partial complete, no information	119	1.6
Partial complete, insufficient information	93	1.2

NOTE: *Partial complete, no information* refers to sampled members who loaded the survey consent form but did not complete any survey questions. *Partial complete, insufficient information* refers to sampled members who answered at least one survey question, but were missing the measure of sexual assault or unwanted sexual contact.

without a response (64 percent) after repeated attempts was higher than in the active-component sample (44 percent). Of the partial respondents, 56 percent provided no information while the rest provided insufficient information to determine whether they had experienced a sexual assault in the past year.

Response Rates

Active Component

In Table A.6, we have used the most conservative of the American Association of Public Opinion Research definitions of response rates (RR1). We present the sample size and number of completes in columns 1 and 2, respectively. Column 3 shows the unweighted response rate, while column 4 displays the design-weighted response rate, with the design weights adjusting for the oversampling of women relative to men. The unweighted and design-weighted versions of the RR1 metric for the active-component Coast Guard sample are 51.6 percent and 50.9 percent, respectively, compared with the overall DoD rate of 30.4 percent. Service-specific response rates in DoD were Air Force (43.5 percent), Army (29.4 percent), Navy (23.3 percent), and Marine Corps (20.6 percent). The response rates for the short, medium, and long forms among Coast Guard service members were 52.1 percent, 52.4 percent, and 50.7 percent, respectively. These small differences are likely due to a difference in the recruitment materials that indicated the length of the survey.

Table A.6
Response Rates by Form Type for the Coast Guard Active Component

	Sample Size	Respondents	Unweighted Response Rate	Weighted Response Rate
Total	14,167	7,307	51.6%	50.9%
Short form	3,961	2,064	52.1%	51.0%
Medium form	3,956	2,074	52.4%	51.9%
Long form	6,250	3,169	50.7%	50.2%

The response rate for women (53.1 percent) was three percentage points higher than that for men (50.5 percent), a smaller difference than that observed for the DoD sample (Table A.7). Across pay grades, senior officers (O4–O6) had a response rate (70.8 percent) almost 30 percentage points higher than that of junior enlisted (E1–E4), who had the lowest response rate (43.3 percent).

Reserve Component

The response rate for the reserve-component sample was 33.4 percent, almost 20 percentage points lower than the 51.6 percent response rate among the active-component Coast Guard (Table A.8). However, this repose rate was higher than the DoD reserve-component response rate of 22.6 percent. The short form response rate was 33.9 percent—comparable to the medium form response rate (33.0 percent).

The response rate for women (38 percent) was higher than that for men (32.5 percent), a larger difference than that observed for the Coast Guard active-component sample (Table A.9). Across pay grades, senior officers (O4–O6) had a response rate (53.2 percent) that was almost 30 percentage points higher than that of junior enlisted (E1–E4), who had the lowest response rate (23.6 percent).

Table A.7
Response Rates for the Coast Guard Active Component, by Gender and Pay Grade

	Sample Size	Respondents	Unweighted Response Rate	Weighted Response Rate
Men	8,315	4,201	50.5%	50.5%
Women	5,852	3,106	53.1%	53.1%
E1–E4	4,937	2,137	43.3%	40.1%
E5–E9	6,625	3,500	52.8%	53.2%
O1–O3	1,638	985	60.1%	59.4%
O4–O6	967	685	70.8%	70.1%

Table A.8
Response Rates for the Coast Guard Reserve, by Form

	Sample Size	Respondents	Unweighted Response Rate	Weighted Response Rate
Total	7,592	2,537	33.4%	33.4%
Short form	3,753	1,272	33.9%	33.9%
Medium form	3,839	1,265	33.0%	33.0%

Table A.9
Response Rates for the Coast Guard Reserve, by Gender and Pay Grade

	Sample Size	Respondents	Unweighted Response Rate	Weighted Response Rate
Men	6,325	2,055	32.5%.	32.5%
Women	1,267	482	38.0%	38.0%
E1–E4	2,488	587	23.6%	23.6%
E5–E9	3,979	1,392	35.0%	35.0%
O1–O3	601	279	46.4%	46.4%
O4–O6	524	279	53.2%	53.2%

Although we did sample all eligible Coast Guard Reserve members, it is worth noting that, because of the much smaller population size and the slightly lower response rate, we had considerably fewer reserve-component respondents ($N = 2,537$) than active-component respondents ($N = 7,307$). This limits our ability to conduct parallel analyses across these two groups, for example, comparing rates of sexual assault by pay grade or by types of sexual assault.

Weighting

After respondents and nonrespondents were identified, we derived survey weights to produce estimates from the respondents' data that are generalizable to the full population of interest. Survey weighting is necessary to make the analytic sample more representative of the population (Heeringa, West, and Berglund, 2010; Little and Rubin, 2002; Schafer and Graham, 2002). Specifically, analyses should incorporate weights that adjust for differential sampling probabilities and nonresponse, and nonresponse weights should "make use of the most relevant data available" to ensure a representative analytic sample (Office of Management and Budget, 2006, Guideline 3.2.12).

Design Weights

For active-component service members, women were selected with certainty (sampling probability of 1) while 25 percent of men were selected for the study. An unweighted average of the respondents' survey reports would not correctly represent population results: it would overrepresent the opinions and experiences of women, relative to their share of the active-component population. Thus, design weights were necessary to adjust estimates for the different sampling probabilities by gender. The design effect, or variance inflation factor, associated with our design is 1.30. (We included everyone in the reserve component.)

Nonresponse Weights

Respondent data were weighted to ensure that our analytic sample was representative of the active-component population. When presenting results for the new assessments from the RMWS forms, we used weights designed to make the analytic sample representative on a broader range of factors than the weights used in 2012 analyses (see Chapter Five of Volume 1 for a detailed description). In the RMWS weighting method, the distribution of the weighted respondents matched the Coast Guard population across key reporting categories of gender and pay grade (Table A.10). The RAND weights included a broader range of factors (see Table 5.2 in Volume 1) than have been included in prior rounds of this survey, to reduce potential nonresponse bias in the survey estimates to the fullest extent possible by including many observed factors.

While including all factors that could plausibly explain nonresponse has advantages for reducing bias, it can have the undesirable effect of making the weights more variable and thereby reduce the precision of estimates. The design effect associated with the RMWS weights among the Coast Guard active-component respondents is

Table A.10
Balance of Weighted Respondents to the Coast Guard Active-Component Population

Reporting Category	Population	Population Percentage	RMWS Weighted Percentage
Female, Coast Guard, Junior Enlisted	2,515	6.43	6.43
Female, Coast Guard, Senior Enlisted	2,047	5.23	5.23
Female, Coast Guard, Junior Officer	900	2.30	2.30
Female, Coast Guard, Senior Officer	390	1.00	1.00
Male, Coast Guard, Junior Enlisted	9,643	24.65	24.65
Male, Coast Guard, Senior Enlisted	18,298	46.78	46.78
Male, Coast Guard, Junior Officer	2,959	7.57	7.57
Male, Coast Guard, Senior Officer	2,360	6.03	6.03
Total	39,112	100.00	100.00

1.9. An estimate of precision is provided by the effective sample size, which was 3,846 (7,307/1.9).

Reserve-Component Weights

The weights for the reserve component were derived through a process that was similar to the RMWS weights for the active-component sample. There were some differences, however, in the process of deriving reserve-component weights. These differences were necessary due to either the nature of the reserve component data or the smaller sample size for those analyses.

First, we had several additional administrative variables for reservists in addition to the variables listed in Volume 1, Table 5.2. This included reserve component, reserve component category designator code, training and retirement category designator, reserve category group code, and days spent on military duties since August 1, 2013. All of these variables were included in the models used to predict key outcomes in the first stage of the derivation of nonresponse weights.

Second, in the initial stage of the development of nonresponse weights, we created variables that captured the relationship between the administrative data (the predictor variables) and key study outcomes. For the active component, we considered six key outcomes, but for the reserve component we only considered three: *sexual harassment*, *gender discrimination*, and *any sexual assault*. Therefore, we derived only three combination variables to be included in the nonresponse model.

Third, in the reserve-component nonresponse model, we created weights that balanced the respondent sample to the full population on the following factors: gender, reserve component (Air National Guard, Air Force Reserve, Army National Guard, Army Reserve, Marine Corps Reserve, Navy Reserve, Coast Guard Reserve), pay grade (E1–E5, E6–E9, O1–O3, O4–O6), form type (short, medium), the three combination variables (created in the prior stage), and all two-way interactions between those seven variables.

Fourth, the reserve-component sample was post-stratified on gender by reserve component as a final step. There were 1,267 women in the Coast Guard Reserve population, corresponding to 16.7 percent of both the population and the weighted respondents; there were 6,325 men in the Coast Guard Reserve population, corresponding to 83.3 percent of both the population and the weighted respondents.

References

American Association for Public Opinion Research, *Standard Definitions: Final Dispositions of Case Codes and Outcome Rates for Surveys*, 7th ed., 2011.

Department of Defense Directive 1350.2, Department of Defense Military Equal Opportunity (MEO) Program, August 18, 1995, Incorporating Change 1, May 7, 1997, Certified Current as of November 21, 2003.

Hanley, James A., and Abby Lippman-Hand, "If Nothing Goes Wrong, Is Everything All Right? Interpreting Zero Numerators," *Journal of the American Medical Association*, Vol. 249, No. 13, 1983, pp. 1743–745.

Heeringa, Steven G., Brady T. West, and Patricia A. Berglund, *Applied Survey Data Analysis*, Boca Raton, Fla.: CRC Press, April 5, 2010.

Kish, Leslie, *Survey Oversampling*, Oxford, England: Wiley, 1965.

Little, Roderick J., and Donald B. Rubin, *Statistical Analysis with Missing Data*, 2nd ed., New York: Wiley-Interscience, 2002.

McMahon, Brian, and Thomas Feger Pugh, *Epidemiology: Principles and Methods*, Boston, Mass.: Little, Brown & Company, 1970.

Morral, Andrew R., Kristie L. Gore, and Terry L. Schell, eds., *Sexual Assault and Sexual Harassment in the U.S. Military: Volume 1. Design of the 2014 RAND Military Workplace Study*, Santa Monica, Calif.: RAND Corporation, RR-870/1-OSD, 2014. As of March 2, 2015:
http://www.rand.org/pubs/research_reports/RR870z1.html

———, *Sexual Assault and Sexual Harassment in the U.S. Military: Volume 2. Estimates for Department of Defense Service Members from the 2014 RAND Military Workplace Study*, Santa Monica, Calif.: RAND Corporation, RR-870/2-OSD, 2015a. As of April 30, 2015:
http://www.rand.org/pubs/research_reports/RR870z2.html

———, *Sexual Assault and Sexual Harassment in the U.S. Military: Annex to Volume 2. Tabular Results from the 2014 RAND Military Workplace Study for Department of Defense Service Members*, Santa Monica, Calif.: RAND Corporation, RR-870/3-OSD, 2015b. As of April 30, 2015:
http://www.rand.org/pubs/research_reports/RR870z3.html

———, *Sexual Assault and Sexual Harassment in the U.S. Military: Annex to Volume 3. Tabular Results from the 2014 RAND Military Workplace Study for Coast Guard Service Members*, Santa Monica, Calif.: RAND Corporation, RR-870/5-USCG, 2015c. As of May 26, 2015:
http://www.rand.org/pubs/research_reports/RR870z5.html

———, *Sexual Assault and Sexual Harassment in the U.S. Military: Volume 4. Investigations of Potential Bias in Estimates from the 2014 RAND Military Workplace Study*, Santa Monica, Calif.: RAND Corporation, RR-870/6-OSD, forthcoming.

National Defense Research Institute, *Sexual Assault and Sexual Harassment in the U.S. Military: Top-Line Estimates for Active-Duty Coast Guard Members from the 2014 RAND Military Workplace Study*, Santa Monica, Calif.: RAND Corporation, RR-944-USCG, 2014. As of March 4, 2015:
http://www.rand.org/pubs/research_reports/RR944.html

———, *Sexual Assault and Sexual Harassment in the U.S. Military: Top-Line Estimates for Active-Duty Service Members from the 2014 RAND Military Workplace Study*, Santa Monica, Calif.: RAND Corporation, RR-870-OSD, 2014. As of March 2, 2015:
http://www.rand.org/pubs/research_reports/RR870.html

Office of Management and Budget, *Standards and Guidelines for Statistical Surveys*, Washington, D.C., September 2006. As of October 3, 2014:
http://www.whitehouse.gov/sites/default/files/omb/inforeg/statpolicy/standards_stat_surveys.pdf

Schafer, Joseph L., and John W. Graham, "Missing Data: Our View of the State of the Art," *Psychological Methods*, Vol. 7, No. 2, 2002, pp. 147–177.

U.S. Coast Guard, *Coast Guard Civil Rights Manual*, Washington, D.C.: U.S. Department of Homeland Security, Commandant Instruction M5350.4C, May 20, 2010.

U.S. Code, Title 10, Armed Forces, Part I, Organization and General Military Powers, Chapter 23, Miscellaneous Studies and Reports, Section 481, Racial and Ethnic Issues, Gender Issues: Surveys. As of March 26, 2015:
http://www.gpo.gov/fdsys/granule/USCODE-2011-title10/USCODE-2011-title10-subtitleA-partI-chap23-sec481